SHAPING MACHINE AN[...]

SHAPING MACHINE
AND
LATHE TOOLS

IAN BRADLEY

MODEL & ALLIED PUBLICATIONS
ARGUS BOOKS LIMITED
14 St. James Road, Watford, Herts.

© Argus Books Ltd and Ian Bradley, 1976

ISBN 0 85242 485 X

All rights reserved. No part of this publication may be reproduced in any form without the prior permission of the publisher.

First published	1949
Reprinted	1954, 1959, 1967 and 1971
This edition, completely revised and re-titled	1976
Reprinted	1981

Made and Printed in Great Britain by Unwin Brothers Limited, Old Woking, Surrey

PREFACE

To produce accurate and well-finished work in the lathe or shaping machine it is essential that the tools used should be of the correct form for the operations undertaken and, besides this, they must be properly sharpened.

An endeavour has, therefore, been made in this book to deal primarily with the practical side of this important subject, in the hope that those interested in improving the quality of their work and in securing greater machining efficiency will find, herein, a little that is both helpful and instructive.

Since this book was first published in 1949 under its original title *Lathe and Shaping Machine Tools*, it has been revised and up-dated with each successive reprint. With this entirely revised and expanded edition the title has been changed to a more explicit form. Apart from minor changes and additions to the text, a 16 page section of additional illustrations has been added in the middle of the book supplementing the diagrams and information given in the main text.

CHAP.	CONTENTS	PAGE

I. MATERIALS FOR MAKING TOOLS   1
Carbon steel—High-speed steel—Stellite—Tungsten carbide—Diamond tools.

II. FORMATION OF THE TOOL'S CUTTING EDGE   8
Clearance angles—Rake angles—Relief angle—Machining : steel, cast iron, brass alloys, aluminium, copper, plastic materials.

III. LATHE TOOLS   16
Tool holders—Forms of lathe tools—Front tool—Knife tool—L-headed tool—V-tool—Parting tool—Backfacing tool—Boring tools—Screw cutting tools—Square threads—Acme threads—Hand turning tools—Tungsten carbide tools—Mounting tools in the lathe—Types of lathe tool holders—Back tool post.

IV. TOOLS FOR USE IN THE SHAPING AND PLANING MACHINE   42
Design of shaping machine tools—Tools for machining steel and cast iron—Tools for brass—Machining aluminium—Tools for special operations—Form tools.

V. SHARPENING LATHE AND SHAPING MACHINE TOOLS   59
Tool wear—Grinding equipment—Honing the tool—Angular grinding—Tungsten carbide tools—Ballbearing grinding machine with cupped wheel.

ADDITIONAL ILLUSTRATIONS
Supplementing text diagrams .. .. i–xvi
(between pages 34–35)

INDEX   73

CHAPTER ONE

MATERIALS FOR MAKING TOOLS

As a result of industrial research many varieties of materials are now available for making the tools used in the lathe, the shaping machine, and the planing machine.

It will be advisable, therefore, at the outset to describe briefly the materials from which these cutting tools are commonly made and, at the same time, to indicate their special properties and the methods employed to prepare them for use.

Carbon Steel. This material, known also as tool steel, was formerly used exclusively for making lathe and shaping machine tools, but it has now been largely superseded by alloy steels with improved cutting qualities. Carbon steel in the soft or annealed state can be readily filed or, when brought to a red heat, it can be hand-forged to make tools of any required form.

Following this, and before use, the tool must be hardened by heating it by means of a blowlamp, gas blowpipe, or electric furnace to a bright red and then plunging it vertically, point downwards, into cold water. Care must be taken not to overheat the steel or to allow the flame to play directly on the cutting edge, for this may so alter the properties of the metal that it is rendered useless for machining purposes.

As the hardening process leaves the steel extremely hard but rather brittle, it must be tempered before use by reheating it to a moderate temperature. To enable the correct degree of temper to be obtained, the forepart of the tool is cleaned with fine emery cloth to form a bright surface, and when gentle heat is applied to the shank, well away from the tip, a play of colours will, after a short while, form on the surface of the steel. As soon as a straw tint reaches the cutting edge, the tool is again plunged into cold water to prevent any further softening of the tip.

If the steel is only tempered to a pale yellow colour it may still be too brittle, but if the blue tint is allowed to reach the cutting edge the tool will be too soft for machining steel.

On completion of the hardening and tempering operations, the tool is made ready for use by forming and sharpening the cutting edges on the grinding wheel. It is important, also, that the under surface of the tool shank should be either filed or ground flat to enable it to lie evenly, and without rocking, on the work table during the grinding process.

Nowadays, carbon steel is but little used for turning tools, except by amateurs, as it so quickly becomes blunted; and if its temper is drawn by overheating during grinding and machining operations, its cutting properties can be restored only by rehardening and temperating. Nevertheless, in the small workshop carbon steel is very useful for making special tools, such as form tools, since it can be so readily filed or machined to shape and then hardened for use.

Carbon steel tools can be purchased either in sets, adapted for carrying out all ordinary turning operations, or as single tools; these have the advantage that they are very much cheaper than the corresponding high-speed steel tools.

Tool steel is usually marketed in the black state in 12 in. lengths of square section of from $\frac{3}{16}$ in. to $\frac{3}{4}$ in. diameter, and also in the form of round rods of from $\frac{3}{8}$ in. diameter upwards.

Silver steel is a variety of carbon steel and has similar properties and hardening qualities, but it has the advantage that it is supplied in the bright state, accurately ground to size, in both the round and square sectional form.

In an emergency, a lathe tool or other small cutting tool can be made from mild steel with the cutting edges case-hardened, but when making cutters in this way it is preferable to use a steel specially manufactured for the purpose as some forms of mild steel do not readily case-harden. After case-hardening, and when the steel is in the fully hard state, the cutting edges will probably be found too brittle for satisfactory use ; the tool should, therefore, be tempered to a straw colour to increase the strength of the cutting portion.

High-speed Steel. This material is an alloy steel containing tungsten and usually other scarce metals in combination ; this greatly increases its resistance to wear and the ill effects of overheating, so that it is capable of working at higher turning speeds and removing greater quantities of material without suffering damage. In addition, this alloy is much stronger than carbon steel and the cutting edge of the tool is not so readily broken by any sudden shock.

In general, this form of steel is hardened by heating it until white-hot and then cooling it rapidly in an air-blast ; but as a rule there is no necessity to employ this method of heat treatment in the small workshop, for the steel can be readily purchased in short lengths in the hardened state and ready for use when ground to the shape required. If, however, the heat treatment of high-speed steel is undertaken in the workshop, a small muffle furnace, heated by gas or electricity and capable of attaining the necessary high temperature, will be found most convenient for the purpose. At the same time, the instructions issued by the steel makers should be closely followed, as these vary for different brands of steel. The air-blast employed for supplying the gas blowpipe can be used to effect the necessary rapid cooling of the heated steel.

The Eclipse brand of high-speed steel suitable for making tools is supplied in the heat-treated state and, in addition, the ends of the square-section material are cut obliquely, leaving but little grinding to be done to bring the tools to shape. The sides of the material are also ground truly flat so that there is no risk of the tool rocking either on the grinding table or in the lathe tool holder. These short lengths of square steel are made in diameters ranging from $\frac{3}{16}$ in. to $\frac{3}{4}$ in. and they will be found eminently suitable for making tools for use in either the tool post or the turret attachment of the small lathe.

This material is also supplied as round cutter bits and in a form specially designed for making parting tools.

Stellite. This material is an alloy composed of cobalt, chromium and tungsten, and it has greater cutting capabilities than high-speed steel, since it is very resistant to wear and the effects of heat during turning operations.

Tools and cutters can readily be made of the solid material, and for large work a tip of this substance can be welded on to a carbon-steel tool shank by means of the oxy-acetylene or electric welding process.

As ordinary Stellite is suitable for cutting steel as well as other materials, a single grade of the substance suffices for all machining operations. Stellite has the additional advantage that it can be ground by employing the methods commonly used for sharpening ordinary high-speed steel tools.

Although the strength of the material is such that a top or side rake of 20 deg. can be employed for machining mild steel, rather rapid wear of the tool point is apt to occur when very light cuts are taken as is usual in the small workshop for turning work to a high finish. On the other hand, Stellite tools give excellent service when used under the more exacting conditions prevailing in commercial engineering.

Tungsten Carbide. This substance is only slightly inferior to the diamond in hardness, but as it is very brittle it is rendered suitable for use as a cutting material by fusing it with metallic cobalt to form a supporting matrix or bond. As tungsten carbide is expensive, it is used in the form of a tip brazed on to a tool shank of ordinary steel to give the necessary support.

The intense hardness of the substance not only greatly reduces tool wear during ordinary turning operations, but it also enables sand castings and even hardened steel to be machined without causing damage to the tool's cutting edge. A further advantage is that the surface speed of the rotating work can be greatly increased and the use of a cutting fluid is unnecessary.

Although the tool's cutting edge will remain sharp for long periods, as compared with a high-speed steel tool, when regrinding becomes necessary this must be carried out on a special wheel, but for the final sharpening operation and for ordinary resharpening the edges are lapped on a wheel impregnated with diamond dust. This process will be fully explained at a later stage when the sharpening of these tools is described in detail.

Although tungsten carbide alloys give exceptionally good service when machining cast-iron and non-ferrous materials, they have an affinity for steel which results in the chips adhering to the surface of the tool ; this, in time, causes pitting or cratering of the tool's surface and the eventual breakdown of the cutting edge.

To overcome this failing, titanium carbide is added to the tungsten carbide alloy used to form the tips of tools intended for machining steel.

Additional strength is given to the cutting edge by reducing the rake angle to some 8 deg., so that for light work the tool does not have the free-cutting slicing action which is a feature of the high-speed steel tool ground with a rake angle of 20 deg. or more.

Owing to the brittleness of tungsten carbide and its lack of strength in withstanding shock or intermittent stress, it is essential, if damage to the cutting edge is to be avoided, that the machines in which these tools are used should be extremely rigid and the work well supported, for any tendency to chatter may result in breakage of the tool point.

When cast iron is machined, the chips separate from the work freely and without greatly stressing the tool's edge, but when turning steel the continuous coiled chips are more tenacious and impose a greater strain on the tool.

Mainly for the above reasons, tungsten carbide tools will be found to give excellent service when machining iron castings but disappointment may be experienced where the equipment is not sufficiently rigid to withstand the more severe stresses connected with machining steel.

The Wimet brand of tungsten carbide tipped tools, manufactured by Messrs. Wickman & Company, is well known, and this firm makes a comprehensive range of tools suitable for machining all classes of materials. The special qualities of these tools are indicated by painting the shanks in various colours, in order to avoid confusion on the part of the machinist and to ensure that the correct type of tool is employed for the work in hand.

In the small machine shop the N grade tool with a red shank should be used for machining cast iron, non-ferrous metals, and other materials in general use. For machining steel, in these circumstances, the manufacturers recommend the S58 grade of tool with a silver-coloured shank; this tool has titanium carbide incorporated with the tungsten carbide tip to enable it to withstand the destructive action resulting from contact with steel already referred to.

These, then, are the materials in common use from which the tools employed for machining are made. To sum up: carbon steel can readily be filed to form any tools of special shape that may be required, and it is then easily hardened and tempered to impart the necessary cutting qualities; high-speed steel is stronger, wears longer and machines more quickly; Stellite has properties intermediate between high-speed steel and tungsten carbide; the latter is expensive but retains its sharpness for long periods and will cut even hard materials at greatly increased speeds.

The figures given in the following table indicate approximately the relative turning speeds that may be employed when using tools made of the materials described. For example, a carbon steel turning tool is represented as operating on mild steel turning at a surface speed of 50 ft. per minute, that is to say, a round bar 1 in. in diameter would revolve at approximately 190 r.p.m.

Tool material	Surface speed in ft. per min.	Revs. per min. for 1 in. diam. bar
Carbon steel	50	190
High-speed steel	100	380
Stellite	125 to 250	475 to 950
Tungsten carbide	250 to 400	950 to 1,520

As an instructive demonstration of the capabilities of these materials they were used in turn to take a facing cut in the lathe across a cast-iron faceplate 9 in. in diameter turning at 80 r.p.m.

When fed outwards from the centre the carbon steel tool failed when a diameter of 2 in. was reached ; the high-speed steel tool cut normally up to a diameter of $4\frac{1}{2}$ in. and then became blunted so that it merely rubbed against the work ; but the tungsten carbide tool cut perfectly and produced a high finish right up to the full diameter of the faceplate.

Diamond Tools. Before leaving the subject of materials used for making turning tools, the diamond-tipped tool should be mentioned as it is now largely used in industry. For example, when aluminium alloy pistons are turned with a diamond tool a superfine finish is produced, and the metal does not ball up on the tip of the tool to form a false cutting edge which would damage the surface of the work ; bearing bushes and the brass parts of clocks are also turned to a high finish in this way. Diamond-tipped tools are also largely used for turning plastic materials, for they operate at high speeds and retain their fine cutting qualities longer than tools made of other substances.

CHAPTER TWO

FORMATION OF THE TOOL'S CUTTING EDGE

THE cutting action of the tool's edge, especially when taking a heavy cut, is essentially a wedging action as represented in Fig. 1 ; and although the more acute the angle of the wedge the easier will be the passage of the tool, this angle cannot be reduced indefinitely without so weakening the tool's edge that it will become unable to withstand the cutting pressure required for a normal rate of machining. Further, the finer the feed the more nearly will this wedging action become a true cutting or shaving process, as illustrated in Fig. 2. Hence, when taking fine cuts, sharpness of the tool's edge, produced by oilstoning, is more than ever essential for ensuring a good finish on the work, although a cutting edge finished by fine grinding will generally be found quite satisfactory for taking roughing cuts.

Clearance Angles. Reference to Fig. 1, which depicts a knife tool in action, will also show that a side clearance angle, as it is termed, of some 10 deg. is necessary to allow the tool to cut into the work.

It will be apparent that the tool, when taking a heavy cut as shown in Fig. 3, advances as though cutting a screw thread or helix on the work, so that the path of the cut lies somewhat obliquely to the transverse axis of the work.

This necessitates the provision of a greater clearance angle than is required for facing a flat surface, for in the latter case a clearance angle of some 5 deg. may suffice to prevent the tool from rubbing against the work.

Rake Angles. When, for example, a knife tool is used for turning mild-steel, the wedge-shaped formation is produced at the tip by grinding its upper surface at an angle of some 20 deg., thus giving it what is called side rake, as represented in Fig. 1. The tool, with a side clearance of 10 deg. and a side rake of 20 deg., now has the form of a wedge of 60 deg. included angle, and the latter is termed the cutting angle.

So far, the tool has been viewed only from its end, and looking at it from the side it will be seen that, as represented in Fig. 4, a front clearance angle of some 10 deg. is given, so that the face of the tool below the cutting edge does not rub against the work and thus interfere with the tool's normal cutting action.

In the tool shown in Fig. 4 it will also be observed that the upper surface is flat in the direction of the tool's long axis, but where, as in a parting tool, for example, the tool is used to

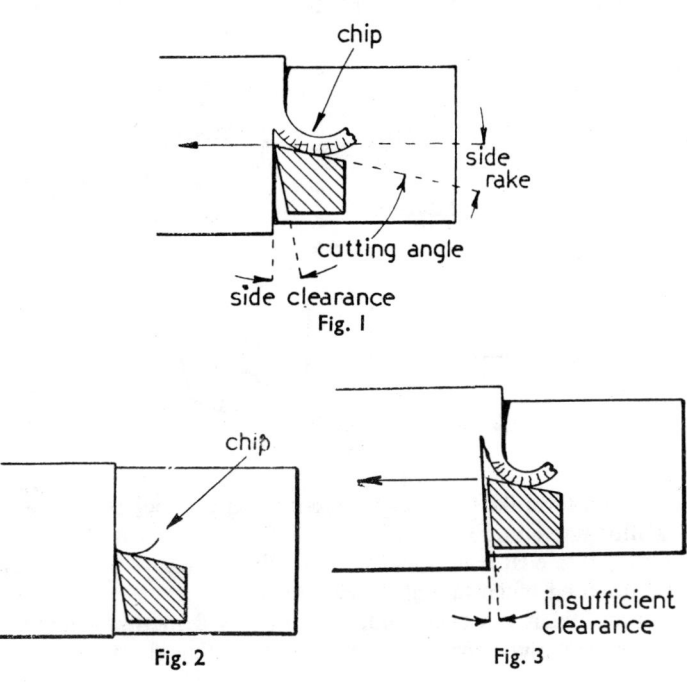

Fig. 1

Fig. 2

Fig. 3

cut inwards towards the centre of the work, the wedge should be formed to lie in this plane as shown in Fig. 5, and the tool is then said to have top or front rake.

Some tools, particularly those used for taking roughing cuts on steel or cast iron, are formed with a combination of both

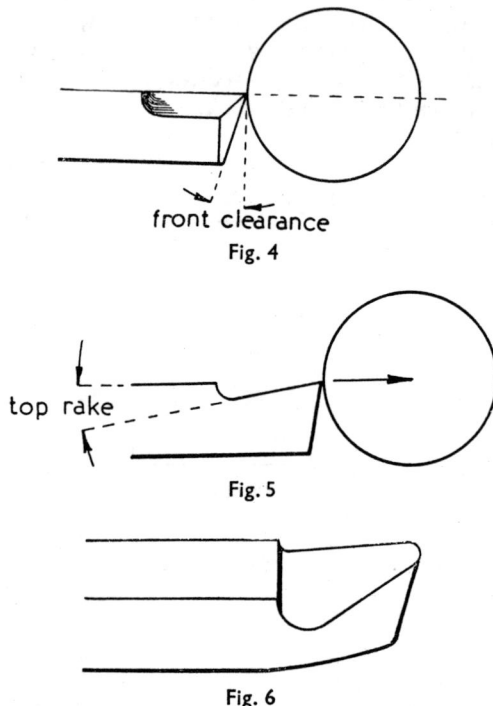

front clearance
Fig. 4

top rake
Fig. 5

Fig. 6

side and top rake to promote free-cutting; a tool of this form is illustrated in Fig. 6, where it will be seen that the tip is rounded to enable it to withstand the increased stress that may be imposed when taking heavy cuts.

From what has been said, it will be evident that clearance angles are given for the purpose of allowing the tool's edge to penetrate into the work and, at the same time, to prevent the face of the tool from rubbing ; rake angles, on the other hand, are provided in order to give the tool's tip the wedge form which facilitates free-cutting by increasing the slicing action of the cutting edge and reducing the friction required to separate the chips.

Relief Angle. Finally, if the tool is viewed from above as represented in Fig. 7, it will be seen that a small portion only

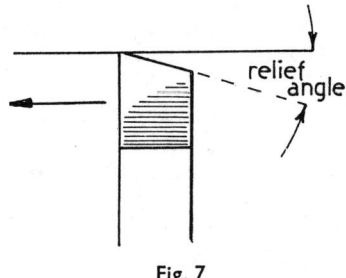

Fig. 7

of the front edge is in contact with the work owing to the relief angle given ; for, if too large a portion of this edge comes into contact with the work, chatter is liable to arise, depending on the rigidity of both the lathe mandrel and the work itself. The relief angle given to a tool of this form should not as a rule exceed some 15 deg., or the tool point will be unduly weakened.

These, then, are the essential angles used when grinding tools to shape, and they can be applied, with modifications where necessary, to all other forms of tools used for machining metals and other materials either in the lathe, the shaping machine or the planing machine.

Now that the general principles governing tool forms have been dealt with, their application to tools used for machining various types of material will next be considered.

Machining Steel. When machining mild steel, and especially the free-cutting varieties, the shavings are formed in long coils and the tool is not necessarily highly stressed even when taking deep roughing cuts. For this reason a large top or side rake angle, or a combination of these, can be used without serious risk of breaking the tool point, and a rake angle of between 20 and 25 deg. is commonly employed where free-cutting and a good work finish are required. The front clearance angle may conveniently be standardised at 10 deg., but it should be borne in mind that excessive clearance may cause chatter besides weakening the tool point. The side clearance when in the direction of the tool's travel may also be fixed at 10 deg. and, as previously mentioned, this should be sufficient to allow the tool to cut freely with all ordinary rates of feed.

Tool steel and alloy steels are much tougher and more difficult to machine than mild steel and, at the same time, the tool point is more heavily stressed. To maintain the strength of the tool's tip, therefore, the front clearance may have to be reduced to some 6 deg., and the side and top rake should not exceed 15 deg. when taking heavy cuts in these materials.

Stainless Steel. Great difficulty is sometimes experienced in machining stainless steel owing to its proneness to work-harden, as it is termed, that is to say, any rubbing action of the tool causes the work surface to become so hard that the machining cannot be continued in the ordinary way. To avoid this occurrence, the clearance angles of the high-speed steel tool used should be ground rather greater than normal, and for the finishing cut the point of the tool should be well rounded; if possible, a single finishing cut only should be taken, as this in itself will probably cause some degree of work-hardening. If, however, a tungsten carbide tool is used to machine this material, no difficulty of machining should arise even if work-hardening does occur.

Cast Iron. When sand castings of this material are machined in the lathe or shaping machine, it is advisable, in order to prevent rapid wear of the tool's cutting edge, to adjust the depth of cut to machine right through the outer scale and into the good metal beneath. For this purpose, the strength of the tool point is maintained and the metal will be cut more freely if a rake of some 12 deg. only is employed in conjunction with the normal clearance angles. Tungsten carbide tools will be found particularly useful for this class of work as their cutting edges are not readily blunted by the sand and scale present; in addition, they will stand up to a fast turning speed, which in conjunction with a fine feed will give a good finish to the work.

Nevertheless, it must be borne in mind that tungsten carbide tools are somewhat easily damaged by sudden shock or intermittent cutting, so care must be taken to avoid these conditions when machining castings of irregular shape; in some cases it may, therefore, be found best to take the preliminary roughing cuts with an ordinary high-speed steel tool and to reserve the tipped tool for the finishing operation.

Brass Alloys. Although the many forms of brasses and bronzes vary greatly both in composition and in machinability, as a rule, the normal clearance angles of 10 deg. should be employed, but no top or side rake is used. Nevertheless, no hard and fast rule can be laid down, for some of these alloys can be machined more readily and to a better finish when the tool is given a few degrees of rake, but if the rake used is excessive the tool is apt to chatter and dig into the work. When difficulty with chatter is experienced, this may, in some cases, be controlled by reducing the front clearance angle of the tool to as little as 5 deg.

Some of the harder bronzes will be found to machine more readily if the tool is given a negative top rake as illustrated in Fig. 8.

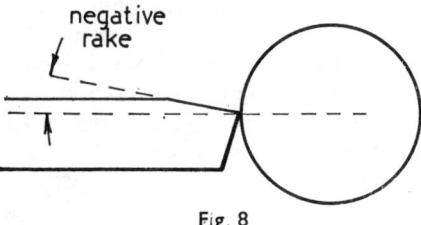

Fig. 8

The successful machining of brass alloys is largely a matter of experiment, as they vary so greatly in their properties, but experience will help to determine the most suitable forms of tools to use and to overcome any difficulties that may arise.

Aluminium. Pure aluminium is seldom used for machine parts, as it lacks strength and is difficult to machine owing to its tendency to ball up on the point of the tool and so cause a ragged surface on the work. On the other hand, aluminium

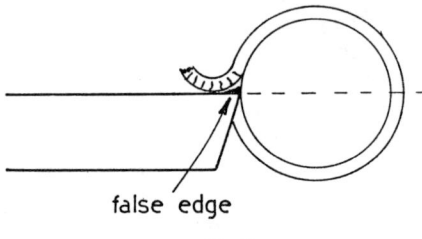

Fig. 9

alloys, such as duralumin, can readily be machined to a good finish if the correct methods are employed. The tool angles used in the case of mild steel will give good results, but the rake angle may with advantage be increased to enhance the slicing effect of the cutting edge, for the strength of the tool can safely be reduced, if required, owing to the very free-cutting properties of this metal.

To produce a good surface finish the point of the tool should be slightly rounded, but if the radius of the tip is too great chatter may result. To prevent the metal from balling up on the upper surface of the tool and forming a false cutting edge, as illustrated in Fig. 9, it is important that the faces forming the cutting edge should be rendered smooth by oilstoning the tip of the tool, as will be described in a later chapter.

Tungsten carbide tools will be found particularly suitable for machining aluminium, for not only are the cutting edges lapped to a fine finish, but the extreme hardness of the surface assists in preventing the softer metal from adhering to the tip.

Copper. It sometimes happens that the commutator of an electric motor or dynamo has to be machined to restore its surface after prolonged use. When a high-speed steel tool is used in the lathe for this purpose, the front and side clearance angles should be increased to 12 and 14 deg. respectively, and ample rake, as in the case of tools used for turning mild steel, should be given.

Tungsten carbide tipped tools made specially for this work have a top rake of some 30 deg., and it will be found that they give a high finish to the work.

In commercial practice diamond-tipped tools are used for finishing commutators and this operation imparts an almost mirror-like finish to the surface.

Plastic Materials. Although these substances vary greatly in their composition, some cause rapid wear of even high-speed steel tools, and for this reason a tungsten carbide tool is to be preferred for any prolonged machining operation.

A tool without top or side rake should be selected for turning these materials, but the side clearance angle can be increased to some 15 deg. with advantage.

For industrial purposes, such as turning insulating bushes for electrical equipment or machining pipe stems, the diamond-tipped tool is now largely used owing to its ability to withstand the abrasive action common to many forms of plastic materials.

CHAPTER THREE

LATHE TOOLS

TOOLS can readily be fashioned to any particular shape required by a process of hand-forging, and Fig. 10 shows two forged tools of a form that was at one time largely used, as but little grinding of the upper cutting face is sufficient to restore its sharpness ; moreover, the tool has a long working life owing to the large reserve of metal in the head. The upper drawing depicts a left-hand tool, that is to say, one designed to cut in a direction towards the tailstock, and the tool below is a round-nosed right-hand tool used for machining in the reverse direction.

Forged tools of this and of all the other forms commonly used for turning, shaping and planing can be obtained in sets of from six to twenty-four tools, made of either carbon or high-speed steel.

In the small workshop, at any rate, these forged tools have been largely displaced in favour of those ground to shape from short lengths of high-speed alloy steel specially prepared for the purpose. As mentioned in Chapter One, these square-section tool bits of the Eclipse and other brands are supplied with all their sides ground flat to facilitate the accurate grinding of the cutting edges, and also to ensure that the tool lies evenly when mounted in the lathe tool post. In addition, as illustrated in Fig. 11, this material can be obtained with the ends formed either to a simple angle, or to a composite angle ground obliquely across the end face. The latter form usually greatly reduces the amount of grinding required to shape the tool for use.

Tool merchants supply this steel in sizes of from $\frac{3}{16}$ in. square section by 2 in. length to $\frac{3}{4}$ in. by 4 in., but for small lathes the 5/16 in. and 3/8 in. material is most generally used for making tools to mount direct in the lathe tool post.

Tool Holders. The smaller sizes of both the square- and round-section tool bits can be mounted in special tool holders of the type illustrated in Fig. 12. These are usually designed to carry the tool with the point raised at an angle of some 15 deg. ; this automatically sets the top rake angle at this figure and so may save any heavy grinding of this portion of

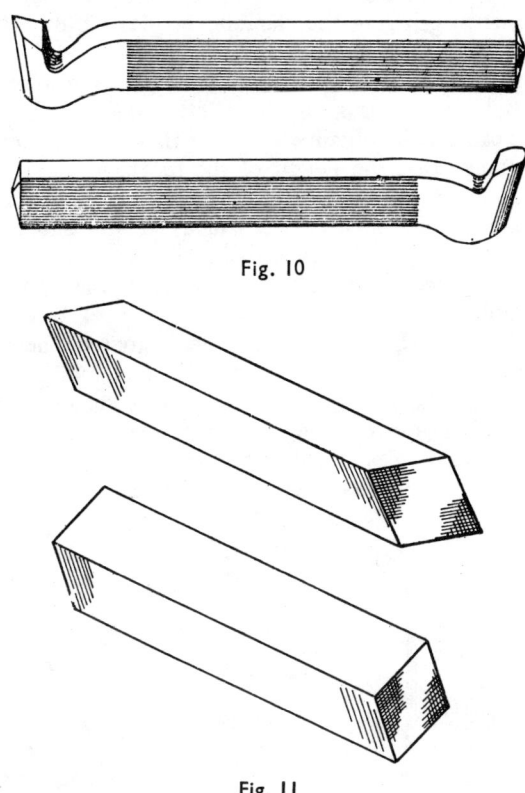

Fig. 10

Fig. 11

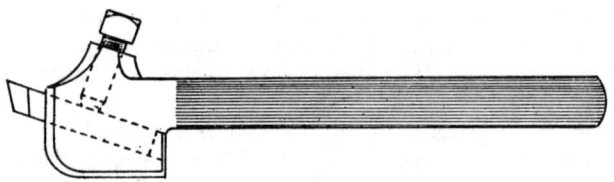

Fig. 12

the tip. Nevertheless, this obliquity must be taken into account when grinding the clearance angles required for the cutting edges.

Although the straight form of holder shown is most generally useful, holders having a right- or left-handed inclination of the head can also be obtained. Whilst these tool holders have the advantage that the height of the tool can readily be set on the inclined clamping surface, they suffer, however, from the disadvantages that the overhang is greater and they are not so rigid as a solid one-piece tool, also the heat engendered during machining is not so rapidly conducted away from the cutting point.

Forms of Lathe Tools. Whether a tool bit held at an inclined angle in a holder is used, or a solid tool is clamped directly in the lathe, the form of the tool's cutting edge and its rake and clearance angles should be identical. Moreover, the general principles already cited apply to the formation of all tools, no matter what their particular purpose.

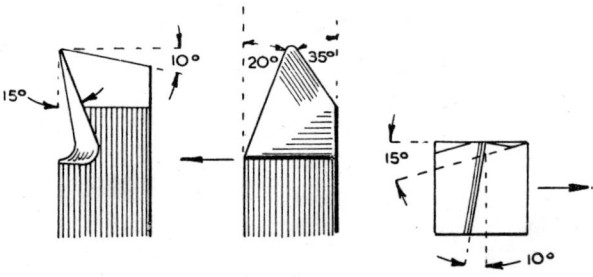

Fig. 13

Front Tool. For taking roughing cuts on mild steel and cast iron the tool shown in Fig. 13 will remove the surplus metal quickly and without unduly stressing the work in any direction. The point is only slightly rounded, and normal clearance angles are used together with a combined side and front rake angle of some 15 deg., ground obliquely in the direction of the shading lines seen at the tip of the tool. When

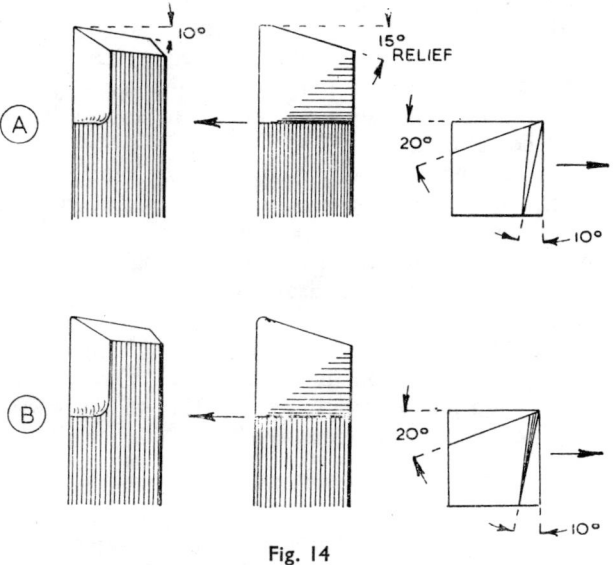

Fig. 14

a tool of this form is used for taking finishing cuts, the cutting point should be rounded to a greater radius to increase the length of the cutting edge in contact with the work. The tool illustrated is of the right-hand form for cutting towards the lathe headstock as indicated by the arrow in the drawing. When taking facing cuts with this tool, a right-hand tool, mounted with its shank parallel with the lathe axis, is used to cut from the centre towards the periphery of the work, whereas a left-hand tool should be fed towards the centre.

With this, as with other forms of tools, the recommendations given for varying the cutting and clearance angles should be observed when machining other types of metals.

Knife Tool. This is undoubtedly the best all-round tool, when made from ground square stock, for general turning operations in the small lathe, and with it both roughing and finishing cuts can be taken for either sliding or surfacing operations.

As shown in Fig. 14, side and front clearance angles of 10 deg. are given, and a side rake of 20 deg. without any front rake is used for turning mild steel; the latter provision ensures that the tool does not lose height appreciably as a result of being resharpened.

In Fig. 14A the front cutting edge is represented as a small flat which will give a good finish to the work and will form a sharp corner up to a shoulder, but this flat must be kept reasonably short or chatter may result during machining.

If a rounded point is used as shown in Fig. 14B, a good finish will also result both when sliding and surfacing with a fine feed, but a rounded corner will be formed when the tool cuts up to a shoulder on the work. A relief angle of some 15 deg. should not be greatly exceeded, as this will weaken the point of the tool unnecessarily.

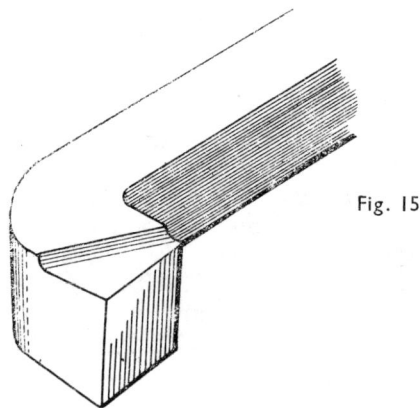

Fig. 15

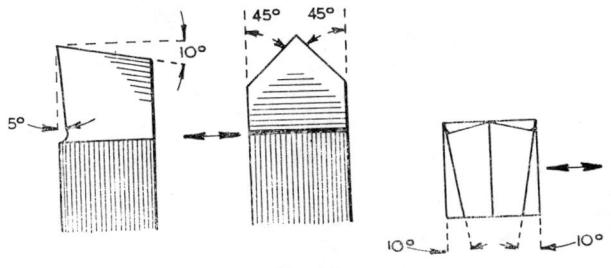

Fig. 16

L-headed Tool. This tool is in many respects similar to the knife tool and it is well adapted for both sliding and surfacing turning operations, but, as shown in Fig. 15, the head of the tool is forged at an angle to the shank to facilitate machining close to the chuck without risk of the chuck jaws fouling the lathe top slide. Although the vertical faces of the tool can be readily ground, its upper surface is not so easily brought into contact with the grinding wheel unless the rake angle is continued right across the upper surface of the end of the shank.

This form of tool is recommended for general use by the makers of tungsten carbide tipped tools where the cutting tip is brazed on to a shank forged with either a right- or left-handed L-shaped head.

V-tool. A tool of the form shown in Fig. 16 will be found useful for chamfering in either direction, for turning grooves to break up a long knurled surface, or for marking-out any required position of the work. If the tip is formed to an included angle of 90 deg., both right- and left-hand chamfers of 45 deg. can be turned, as when finishing nuts and washers, or chamfering both edges of a collar turned on the work. To promote free-cutting, a top rake of from 5 deg. to 10 deg. is used, but no side rake is given, as both sides of the V point are employed as cutting edges.

It is usually advisable to remove the extreme sharp point of the tool on the oilstone in order to strengthen the tip.

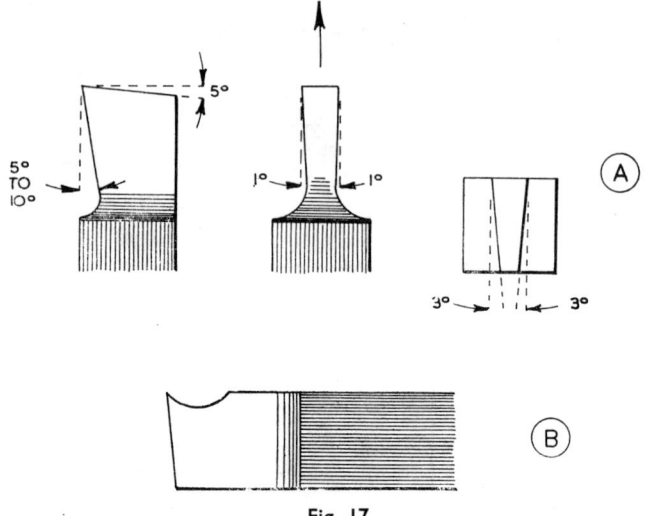

Fig. 17

Parting Tool. In order to lessen the stresses when parting off, and at the same time to avoid chatter, the cutting edge of the tool should be made as narrow as possible consistent with adequate strength.

The clearance angles are also reduced to a minimum to avoid weakening the support given to the cutting edge; 5 deg. will usually be found sufficient for the front clearance angle and 3 deg. at the two side surfaces of the tool. In addition to these clearances, relief is given behind the cutting edge to prevent the tool from jamming in the cut. Although no top rake is used when machining brass, a rake angle of 5 deg. to 10 deg. should be used for parting steel. This top rake should be formed as shown in Fig. 17A, so that the chips do not coil appreciably until they reach the curved surface well away from the tip; for if the rake is ground as in Fig. 17B, the chips will form a tight coil within the groove cut in the work where they are apt to jam and possibly cause breakage of the tool.

The Burnerd patent parting tool illustrated in Fig. 18 is designed to prevent the work from rising under the influence of the upward thrust imparted by the tool. For this purpose, an adjustable thrust piece is fitted to the holder which accommodates a cutter bit having the correct side clearance angles ready-ground.

Although this tool serves its purpose well, the method of using the parting tool upside-down in a back tool post at the rear of the cross slide is now gaining in favour, as it largely overcomes the difficulties experienced when parting off operations are undertaken. The back tool post is discussed in greater detail on pages 40–41.

Back-facing Tool. When it is required, at one setting, to face the inner end of a piece of bored work in addition to its outer end, a back-facing tool can be used of the form shown in Fig. 19, where the tool is depicted facing the inner end of a bush which has already been bored while mounted in the chuck.

Boring Tools. These tools are made in several patterns either in the form of a one-piece tool or as a cutter bit mounted in a suitable holder ; in either case, the form of the cutting edges at the tip is essentially similar. Fig. 20 depicts a solid boring tool with a composite rake angle adapted for machining steel, and with sufficient clearance below the cutting edge, as

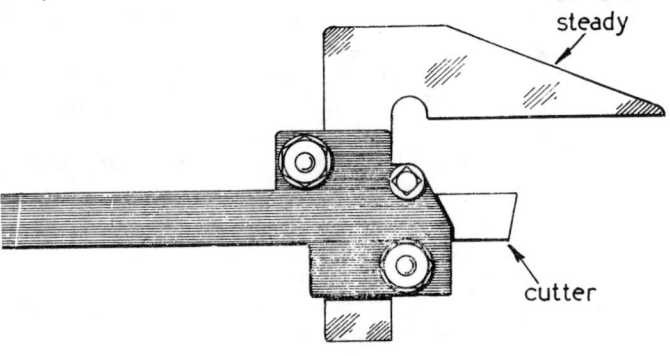

Fig. 18

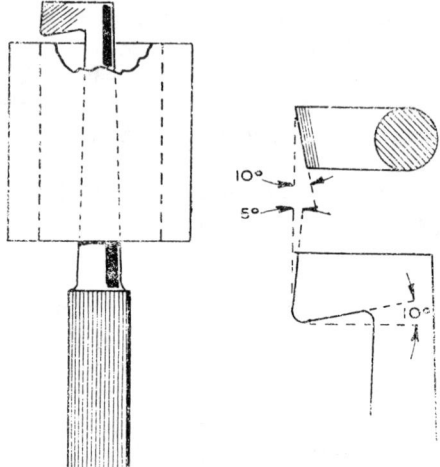

Fig. 19

shown in Fig. 20B, to allow for boring small holes without the risk of the heel of the tool rubbing in the bore.

Tools used for boring brass should be formed with similar clearance angles but with little or no rake.

When taking heavy cuts it is advisable to form the tip of the tool as shown in Fig. 20C, for if the tool is over-stressed it will then tend to be forced away from the work and, moreover, this shape gives added strength to the cutting edge.

It will be clear that a small tool of this type will, in time, become worn out and useless as a result of being reground ; and, to overcome this wastage, a cutter bit mounted in a holder can be substituted and replaced when necessary at small cost.

The Eclipse high-speed steel boring bar bits illustrated in Fig. 21 are obtainable in diameters ranging from $\frac{3}{16}$ in. to $\frac{3}{8}$ in. and with the cutting point ground ready for use. These cutters have a flat formed on their under surface, to enable them to be clamped in the lathe tool holder, but it is the usual practice to mount them in a holder of either of the patterns shown in Fig. 22 so that they can then be firmly secured in the lathe tool holder. Another very convenient type of boring tool is the Nulok boring bar fitted with a round-section tool bit and secured in the lathe tool holder by means of a split carrier of sqaure-sectional form as illustrated in Fig. 23.

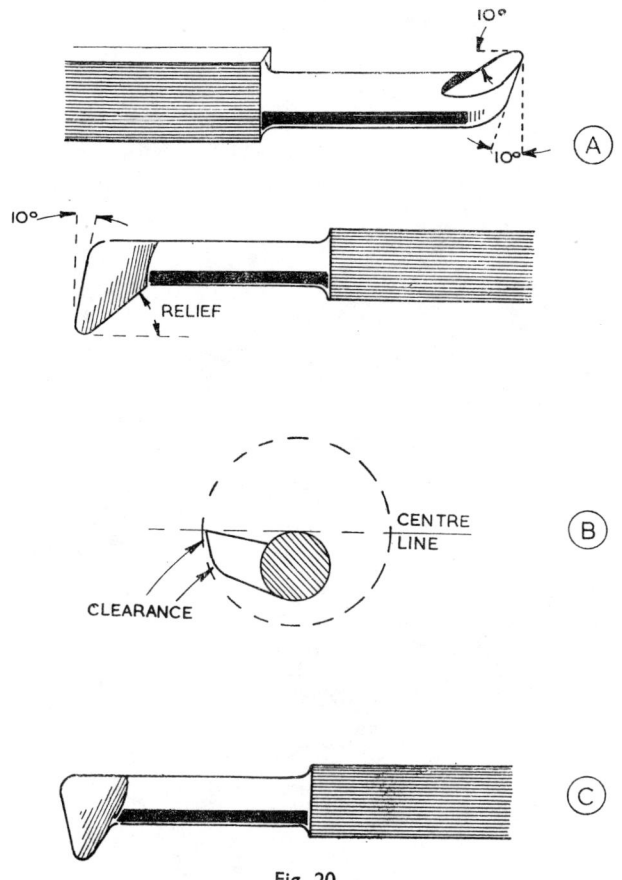

Fig. 20

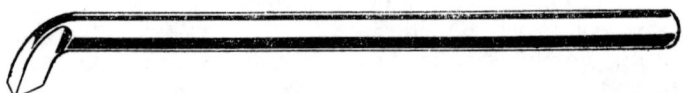

Fig. 21

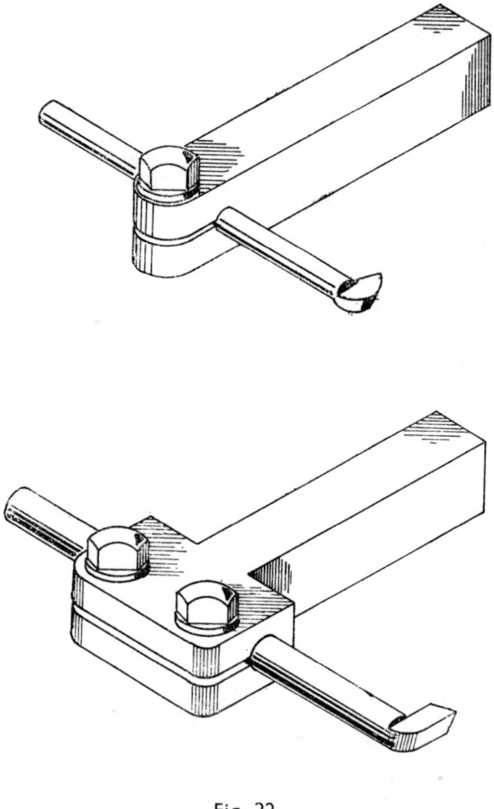

Fig. 22

The end of the bar is cross-drilled in two directions to enable the tool bit to be held either at right angles to the bar or at an oblique angle across it.

The small tool bits used in the bar should, of course, have their cutting edges ground to essentially the same form as that shown for the ordinary solid boring tool; similarly, cutter bits, fitted to boring bars mounted between the lathe centres for boring components secured to the saddle, should be shaped in accordance with the same general plan.

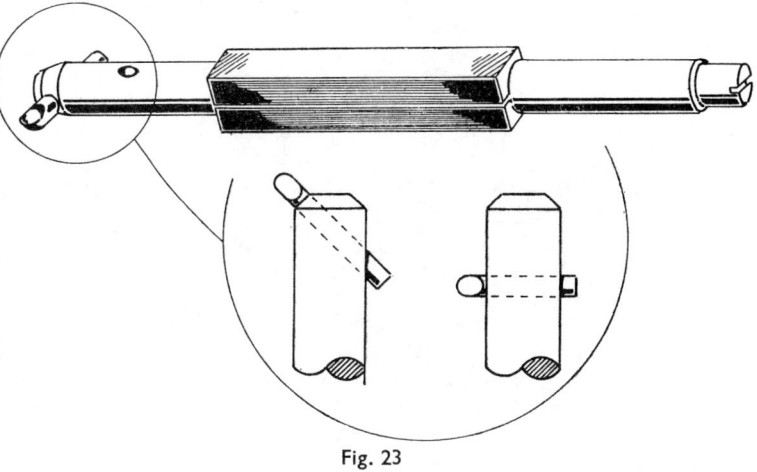

Fig. 23

Screw-cutting Tools. When cutting Whitworth-form external threads, which have an included angle of 55 deg., the point of the tool is also formed to this angle; but, in addition and as shown in Fig. 24, the clearance angle below the leading cutting edge must be rather greater than that in an ordinary V-pointed tool, in order to allow for the obliquity of the cut due to the helix angle of the spiral-form thread. A section of the tool point on the line AB is shown in the lower drawing which represents the tool in operation. The point of the tool will be

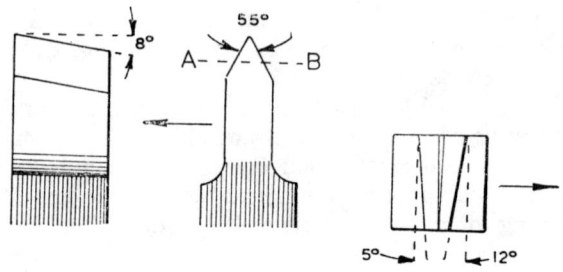

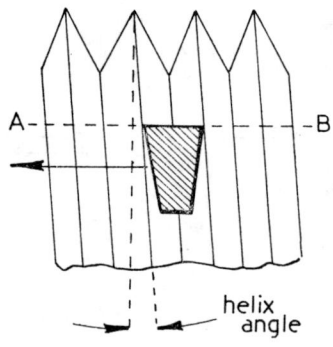

Fig. 24

Fig. 25

strengthened if it is slightly rounded at the extreme tip, and this will also assist in producing the true thread form with its curved root.

Where the method of setting over the lathe top slide to an angle of $27\frac{1}{2}$ deg. for cutting a 55 deg. thread is adopted, a small amount of side rake given to the leading cutting edge will facilitate the machining, but no top rake is used, as this would alter the actual form of the thread cut.

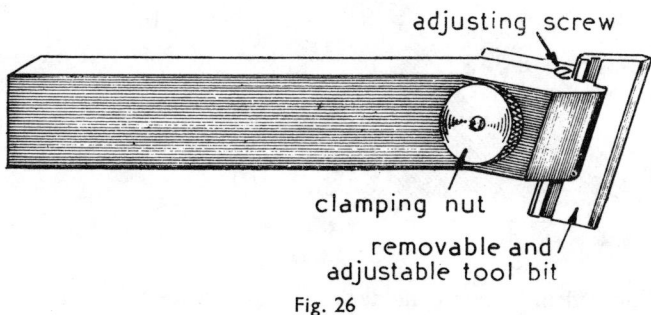

Fig. 26

Although an ordinary one-piece V-tool is commonly used for screw cutting, special tools are obtainable for this purpose.

Fig. 25 illustrates a tool that will cut the correct thread angle and which requires resharpening solely on its upper face.

In this instance, it will be seen that a small amount of top rake is provided to promote free-cutting, but as this reduces somewhat the value of the angle cut in the work, the angle of the tool's point is correspondingly increased to make good the loss.

The simple V-pointed tool cannot form the rounded crests on the threads which is a feature of the Whitworth system; for this purpose a special form-tool is necessary and a separate cutter is required for each pitch of thread. A screw-cutting form-tool of this type, manufactured by Messrs. Jones and Shipman, is illustrated in Fig. 26, and here it is shown

Fig. 27

mounted in a special holder which is provided with a screw mechanism for adjusting the height of the cutter. As reference to Fig. 27 will show, this form of cutter will machine the correct curvature at both the root and the crest of the thread.

For cutting internal threads, either a one-piece tool, as shown in Fig. 28, or a cutter mounted in a holder can be

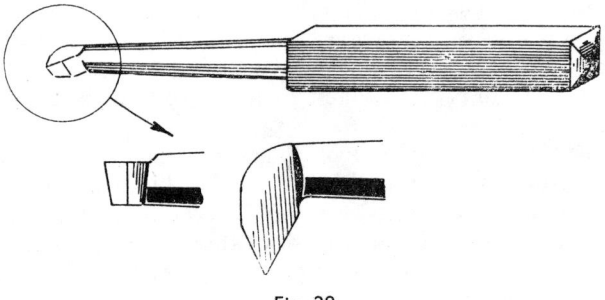

Fig. 28

employed; in either case the tool point is formed like a boring tool, but the V tip must be ground to the correct angle of 55 deg. for the Whitworth thread and, as with the external threading tool, sufficient side clearance must be given to the leading cutting edge to allow for the helix angle of the thread.

The Nulok boring bar, already described, makes a convenient holder for a cutter bit when cutting internal threads. Messrs. Jones and Shipman, however, make special tools to facilitate the machining of internal threads, and that illustrated in Fig. 29 consists of a V-form circular cutter attached to a

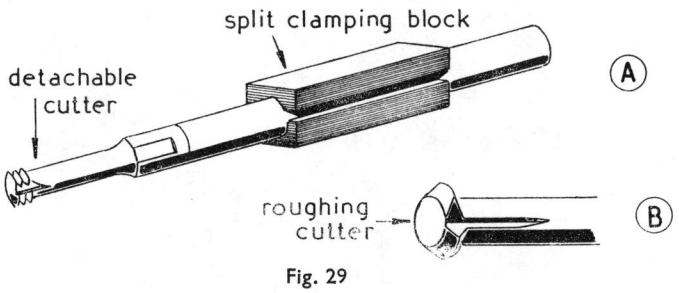

Fig. 29

bar which is secured in the lathe tool holder by means of a split square-section clamping piece. The cutter shown in Fig. 29B is intended for roughing out the thread, and as it has a single V point it may be used to cut threads of any pitch. For finishing the thread to the correct form a cutter of the type shown in Fig. 29A is screwed on to the bar, and as this is a form-tool for shaping the roots and crests of the threads and has more than one cutting point, each cutter can be used for only one pitch of thread. When the Jones and Shipman external threading tool is used in conjunction with its internal counterpart, very accurately fitting and highly-finished threads can be produced.

Square Threads. Small square threads can be cut to the full depth by means of a tool formed like a parting tool, but,

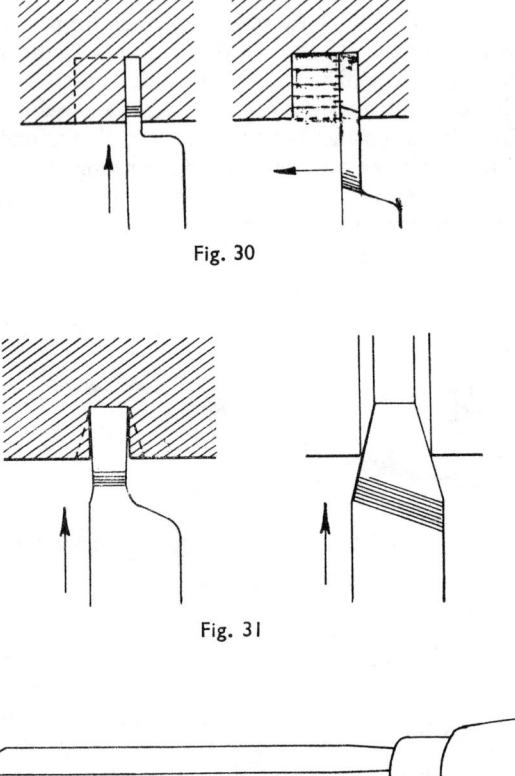

Fig. 30

Fig. 31

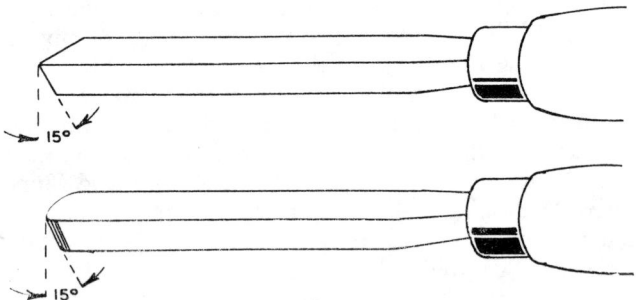

Fig. 32

again, additional clearance must be given to the leading cutting edge in accordance with the helix angle or obliquity of the thread cut.

Large square threads are usually first cut to the full depth with a narrow-edged tool as in the preceding instance; the cut so made is then enlarged to the full width by means of a narrow right-hand knife tool as represented in Fig. 30.

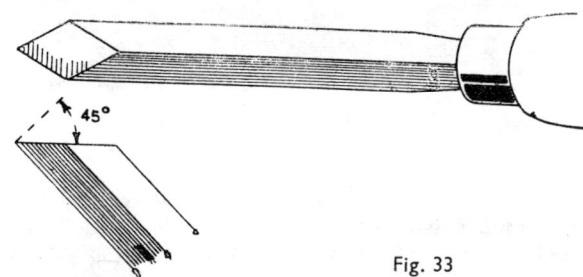

Fig. 33

Acme Threads. Acme threads are formed in the same way by taking a preliminary cut to the full depth with a narrow-edged tool, and then cutting the flanks of the thread to the correct angle of 29 deg. with a form-tool as shown in Fig. 31.

Internal square and Acme threads can both be machined with tools similar to those used for internal V-threading, but the shape of the tool must be adapted to cut the correct thread form so that the counterpart of the external thread is produced.

Hand Turning Tools. These tools of the forms shown in Fig. 32 are mostly used in the small workshop for making or finishing brass parts. As a rule, ample front clearance should be given to allow of the tool handle being depressed when top rake is required to promote easier cutting or a better finish on the work.

The graver, illustrated in Fig. 33, can be used for a variety of turning operations according to the position in which it is applied to the work. The lozenge-shaped flat which is ground at the tip to make the cutting edges is usually formed at an angle of 45 deg. to the shank of the tool.

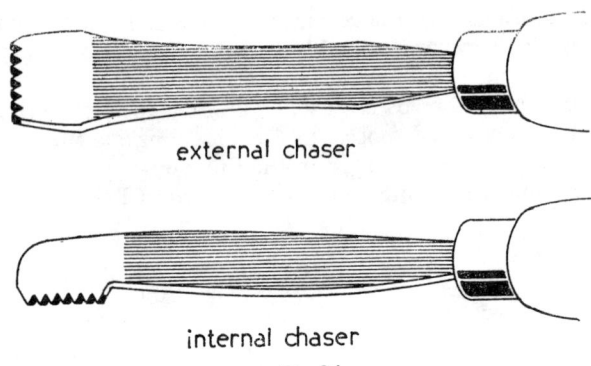

external chaser

internal chaser

Fig. 34

When finishing a thread after it has been cut with a single-point V tool, a hand-chaser is sometimes used and is then supported on the lathe hand rest. These tools, as depicted in Fig. 34, are made in both the external and internal form, but, as in the case of the screw-cutting form-tool, a separate chaser is required for each pitch of thread.

Tungsten Carbide Tipped Tools. So far in this chapter but little mention has been made of tools tipped with tungsten carbide.

These tools are supplied with their cutting edges finely finished and ready for use, whilst the form of the cutting edges is in accordance with the manufacturers' wide experience of what is most suitable to ensure satisfactory service.

As this material is brittle, the cutting edges have to be shaped so that they are well supported by the underlying metal, and at the same time the cutting angles are made rather more obtuse than in the ordinary high-speed steel tools.

Mounting Tools in the Lathe. As it is essential that tools, when in operation, should be supported as near to the cutting point as possible, the importance of forming the tool with a flat under-surface must always be borne in mind. The effect of mounting a tool with an irregularly shaped shank is illustrated in an exaggerated form in Fig. 35 ; here, it will be

(continued beyond page xvi)

Four forged tools. The two on the left correspond with those depicted in Fig. 10 (page 17) while those on the right are of a form once used as finishing tools for turned work. As may be expected they are only intended for taking very fine cuts.

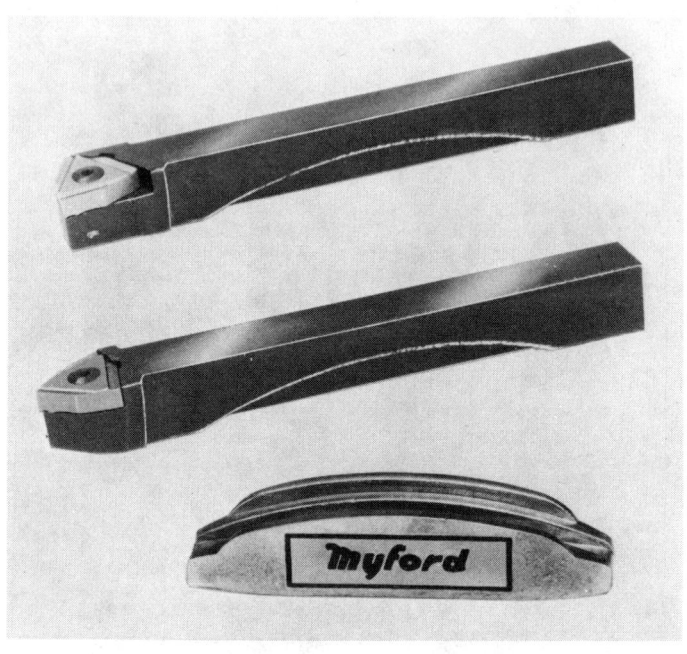

For some years now, Myford of Nottingham, England, have been marketing a range of tools designed to dispense with the pieces of packing that are commonly employed in order to bring the tool point to centre height. To this end the shank of the tool is machined so that it may drop accurately into the "boat" provided for it. As may be observed in the illustration above, the seating for the tool is curved; consequently the point of the tool may be elevated or depressed by adjusting the tool shank in relation to the "boat". The method of making the adjustment is depicted in the diagrams on the opposite page, where the small captions are self-explanatory. This is additional to methods discussed on page 21.

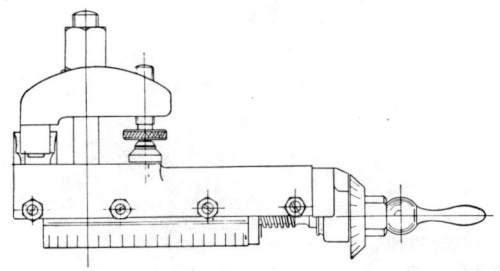

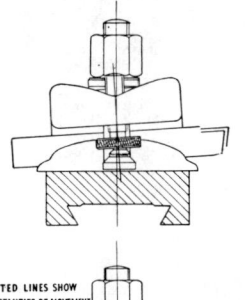

1. Tool clamped in set position, note the large area of base contact.

2. Tool set at high limit illustrating angular movement of tool clamp with square position of adjusting screw loose heel.

3. Range of tool adjustment. The tool boat always remains in firm contact with its base support, clamping being evenly distributed over the whole area of clamping surfaces.

4. Tool shown in extreme low limit setting. This setting can be obtained by a single piece of packing placed under the tool boat for hard brass turning where tool top rake is not required. Note the actions of spherical washer and large radial clamping surface of tool.

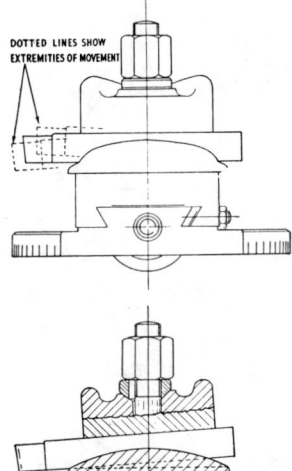

DOTTED LINES SHOW EXTREMITIES OF MOVEMENT

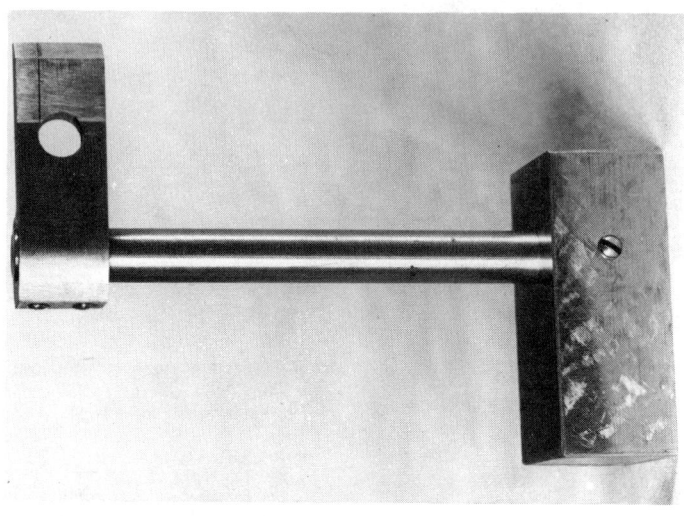

A Tool-setting Gauge

It should be noted that an essential pre-requisite to correct machining is that the point of the turning tool should engage the exact centre-line of the work. In the past there have been many devices for ensuring this requirement, one of which is depicted in the illustration here. This tool-setting gauge consists of a machine base provided with a pillar supporting the blade on which is engraved a line indicating the centre-line. In use the point of the tool is aligned with this mark as depicted in the drawing after the blade has been set with reference to a centre placed in the lathe toolstock. A device of this type can either be mounted on the cross slide or on the lathe bed, whichever is most convenient.

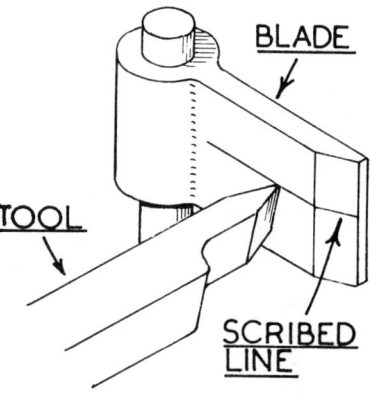

Ball Turning Equipment

Apart from the use of the hand graver for the turning of ball-shaped parts on selected components, equipment for the purpose is for the most part, based on the provision of a tool mounting that can be swung around a central point. In this way the cutting edge of the tool can be made to describe a controlled arc around the work, the radius of this arc being capable of adjustment to suit the work in hand.

The central point around which the tool is swung must, of course, be set on the centre line of the lathe so provision needs to be made to check the alignment, for the most part while a centre set is the lathe tailstock.

Equipment for ball turning varies from the simple to the more advanced, and it is an example of the former type that is illustrated in the picture above and diagram overleaf.

As may be inferred the fixture comprises a sole plate (A) upon which a hand controlled quadrant (B) is mounted. The quadrant, an iron casting, is furnished with a mount for the turning tool which may be adjusted to the radius of the ball it is desired to machine, and then secured by the pair of Allen set screws seen at the top of the casting.

The radius required is measured from the tip of the turning tool to a reference point (C) set in the centre of the spigot around which the quadrant turns, the reference point being of such a height that it can be brought into alignment with the lathe centre, as previously mentioned, so that the axis of the spigot can be set on the centre line of the lathe. The reference point is of course demountable and must be removed before turning can be commenced. Once the alignment has been secured a note of the cross-slide index reading is taken to ensure that turning is only carried on up to this point.

The device illustrated has a maximum capacity of 3 in. dia. and is here seen turning 2 in. spherical components that are anchorages for the canopy supports of a Showman's Traction Engine.

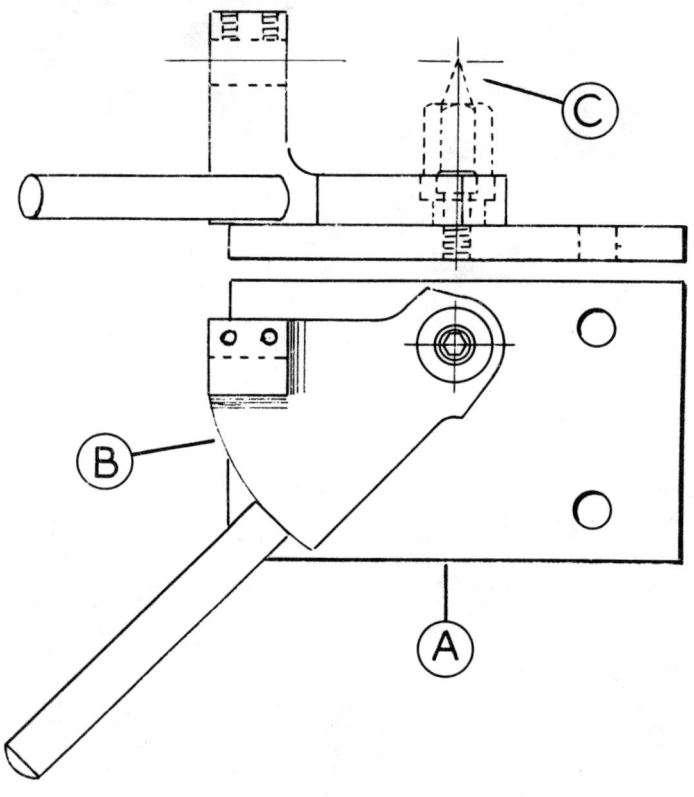

The Dickson Quickchange Toolpost

In the past there have been a number of devices designed to allow the operator to change tools on his lathe rapidly, but without sacrificing the accuracy of tool setting that is essential to efficient working. The latest of these devices has been introduced by Dicksons (Engineering) Ltd. of The Trading Estate, Farnham, Surrey, and an example of the equipment is seen above mounted on the top slide of the Myford Super Seven lathe. The several components of this equipment may be examined in the second picture (next page). In the centre is the toolpost itself; this is passed over the stud fitted to the lathe top slide. The several tool boxes supplied may be seen to either side of the toolpost. The parting tool holder is seen at (1). A holder for round-shank boring tools at (2) whilst at (3) is the tool box provided for square-shank tools and toolbits:

The back of a toolbox may be observed at (4) and from this it will be seen that the box registers with the toolpost on angular ways. It is secured by the eccentrically-operated pad seen at the centre of the post. The eccentric is formed on the hexagon-head spindle which also carries the interrupted register. This register engages the adjustable stops provided on each toolholder. By this means, once the tools have been set to centre height the toolholders may be changed with repeatable accuracy of a high order.

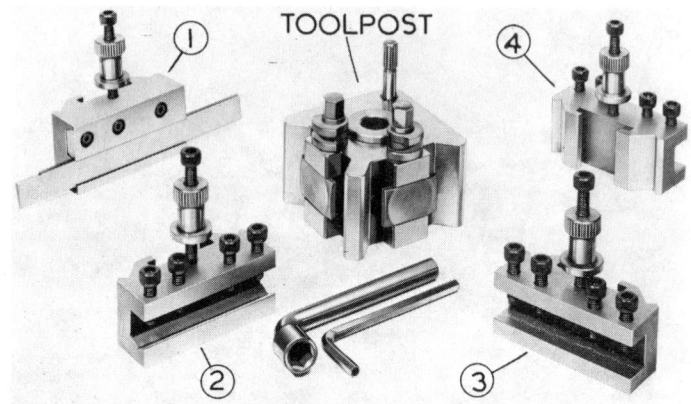

Components of the Dickinson Quickchange Toolpost, as described on the previous page.

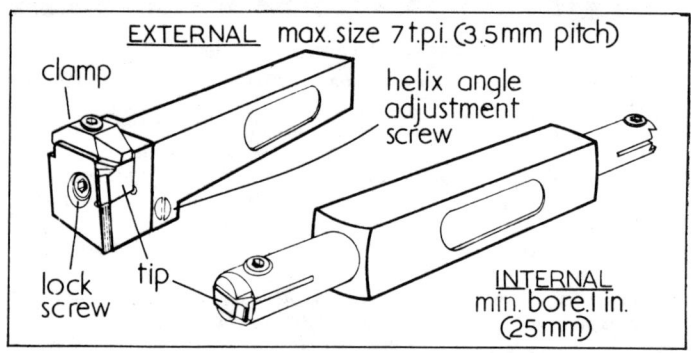

Clamp-tip tools

At a time when clamp-tip tungsten carbide tools are gaining favour, the "Kennet" screw-cutting tool made by Sanderson & Cóster of Highclere, Newbury, Berks, England, is of interest. These tools, as shown above, are available for both internal and external work and are fitted with means of adjusting the tip to correspond with the helix angle subtended by the pitch of the screw thread to be cut and the

The Toolpost illustrated is one made by the author for service on the Myford ML7 lathe. It was constructed in accordance with the drawing depicted in Fig. 43 on page 41.

---

diameter of the work involved. In the case of the external tool, this adjustment is made by turning the screw low down on the side of the screw. At its opposite end then this screw carries a small dial upon which are engraved divisions that need to be used to set the helix angle in accordance with a table supplied by the maker. On the internal tool there are two tool bars provided stamped at the ends A, B, C, D, each end holding the carbide insert at a slightly different angle. The inserts are intended for high speed screw cutting on a lathe that has the necessary rigidity and requisite high spindle speeds. The surface speeds envisaged vary from 300–450 surface feet/minute up to 800 for brass and light alloy.

This Toolpost is one made by the author for service on the Drummond lathe. For the type of construction see Fig. 43, page 41.

The advantage of the Back Toolpost has already been stressed in the text. A good example of this may be seen here where the V-groove in a steel pulley is seen being formed by a direct plunging out. See text on page 40.

Tool with a revolving head as illustrated and described in Fig. 50, page 46.

The two tools illustrated above and on the next page are somewhat unusual. They were produced in order to slit some collets that had been made to hold square material for turning. The work was carried out quite successfully as may be gathered from an examination of the collets which appear in the picture. See also page 58. The slitting was actually made with the smaller of the two tools shown. The success attained with this device led to the author making the larger saw tool depicted on the next page.

The larger of the two slitting (or saw) tools described on the previous page and made by the author.

An example of an internal keyway in use as described on pages 55 and 57.

The internal keyway tool as used in the arrangement shown on the previous page.

seen that the outer point of support is at A, and not at B as it should be, thus making the overhang the unnecessarily long distance from A to C. This form of faulty mounting may be even more serious when a type of lathe or shaping machine tool holder with only a single clamping bolt is used.

No matter how well the tool may be designed for its purpose, it will not operate to the full advantage unless correctly mounted in the machine.

Fig. 35

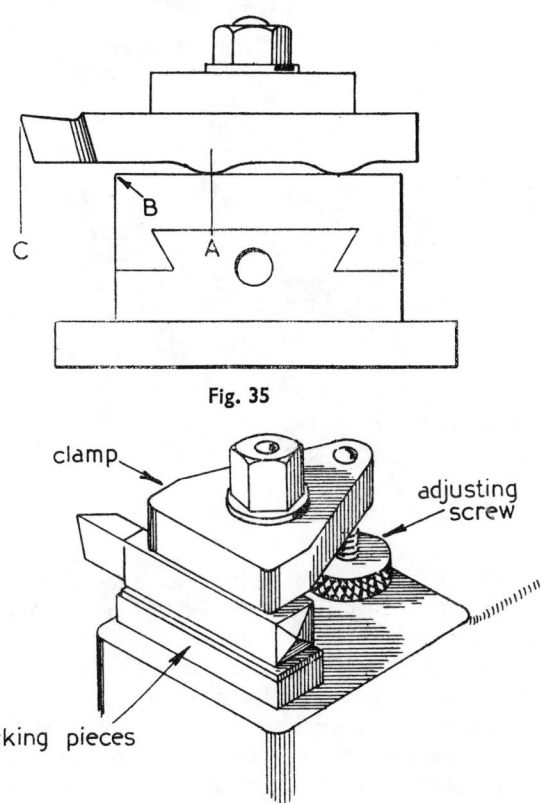

Fig. 36

Types of Lathe Tool Holders. The so-called English type of tool holder illustrated in Fig. 36, although quick in operation, suffers from the disadvantage that it lacks any form of height adjustment, and the tool has to be set to the correct height by means of loose packing strips ; further, unless the top slide is of sufficiently robust construction, undue tightening of the central bolt may distort the slide itself and prevent it from working freely.

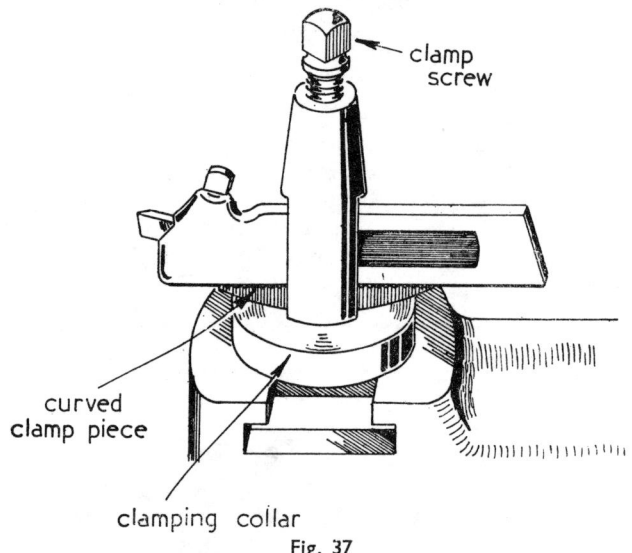

Fig. 37

The American form of tool holder shown in Fig. 37 is provided with a curved, sliding packing piece which allows the height of the tool to be readily adjusted and the tool itself to be secured in place by tightening the central clamping screw. Unfortunately, movement of the packing piece in its curved slot alters the angle at which the tool lies, so that, as shown in Fig. 38, both the front clearance angle and the top rake angle in relation to the work are also upset. It is advisable, therefore, when using a tool holder of this pattern, to employ

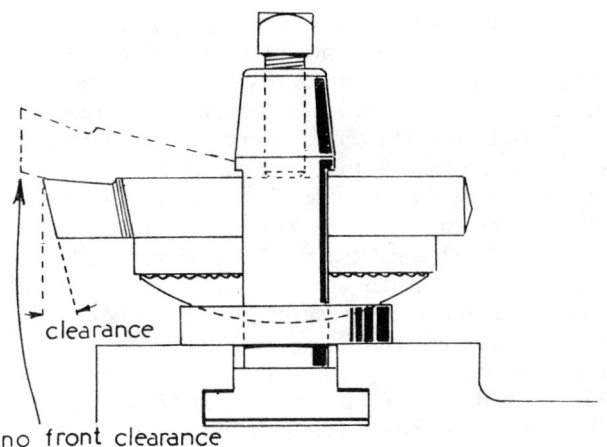

Fig. 38

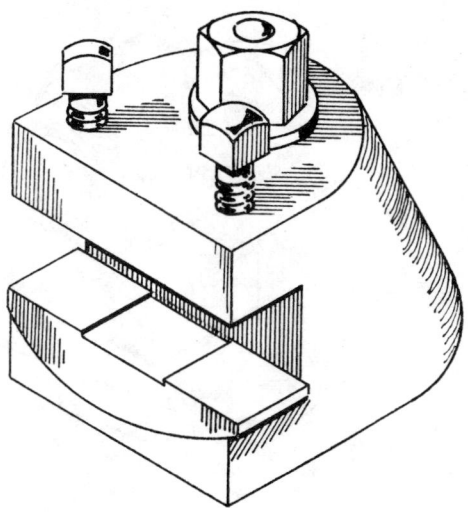

Fig. 39

packing strips for setting the tool roughly to height, and then to make the final adjustment with the aid of the curved slider. This type of tool holder is not so rigid as the English pattern, and for this reason some manufacturers do not recommend it for use with tungsten carbide tipped tools where rigidity is essential to protect the tool from damage.

Distortion of the top slide when clamping the tool can be prevented by employing the box-form of tool holder shown in Fig. 39, which illustrates the Boley pattern holder supplied for this firm's precision lathes. This holder is secured to the T-slot in the top slide by means of a central bolt, or in other lathes the tool post bolt is used for this purpose. The tool is clamped in place by tightening the two clamp screws shown, and a curved, sliding packing piece is fitted to provide for adjusting the tool height over a small range.

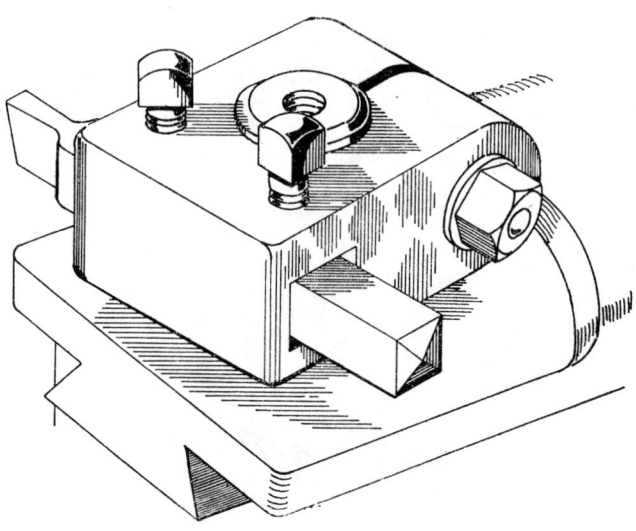

Fig. 40

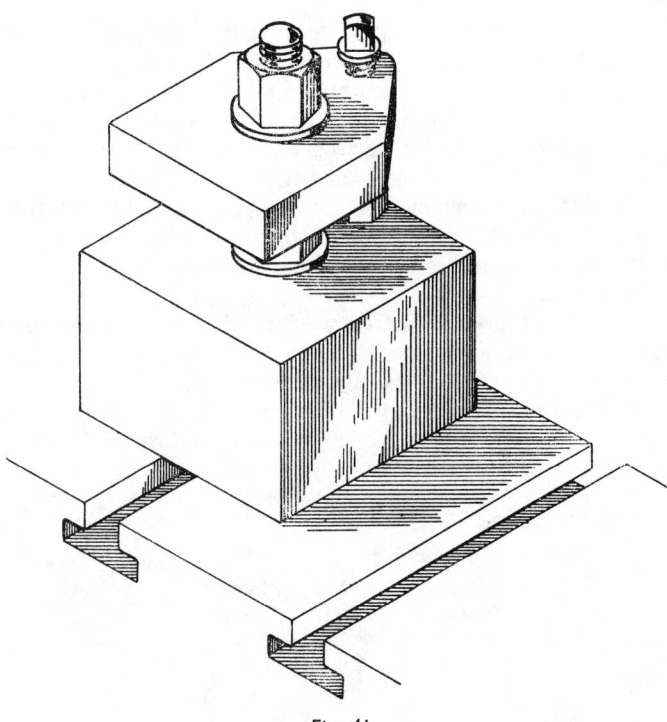

Fig. 41

The well-known Drummond pattern tool holder, illustrated in Fig. 40, clamps to the tool post which is cast in one piece with the body of the top slide. This arrangement not only affords an excellent means of height adjustment, but it also relieves the top slide of all clamping strain.

Where maximum rigidity is required for heavy turning operations, some mechanics prefer to dispense with the top slide by mounting the tool holder directly on the cross slide. For this purpose, an attachment of the form shown in Fig. 41 will be found convenient, as it is secured in place by means of a single central bolt, engaging in one of the T-slots of the cross slide, and the English pattern tool clamp enables tools and tool holders of all sizes to be firmly held.

The four-tool turret illustrated in Fig. 42 is designed to allow any of the tools to be brought into operation as required and, at the same time, to be accurately located by means of a ratchet device to facilitate repetition work. The tools are adjusted to centre height by the use of packing strips, but this is not inconvenient as the tools remain permanently in place in the turret until their removal becomes necessary for resharpening.

Back Tool Post. The advantages of mounting the tool upside-down at the back of the lathe for parting and chamfering operations are now widely recognised, and the back tool post illustrated in Fig. 43 has been specially designed for this purpose. As will be seen, the device is bolted to the T-slots.

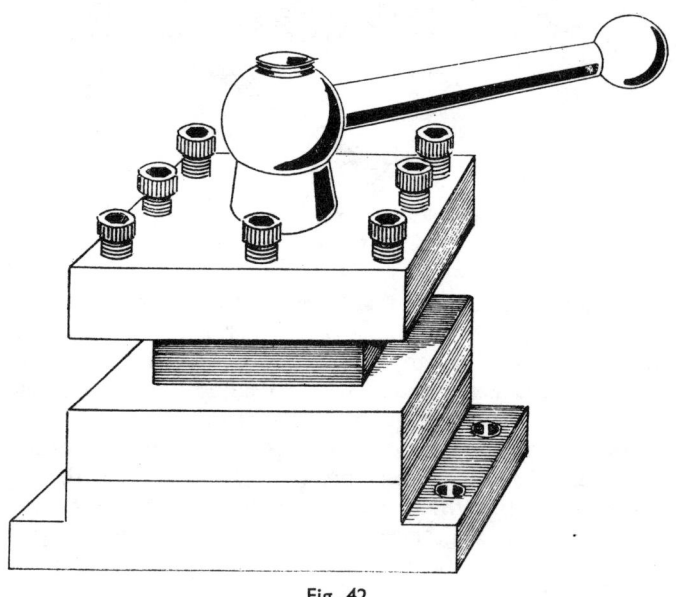

Fig. 42

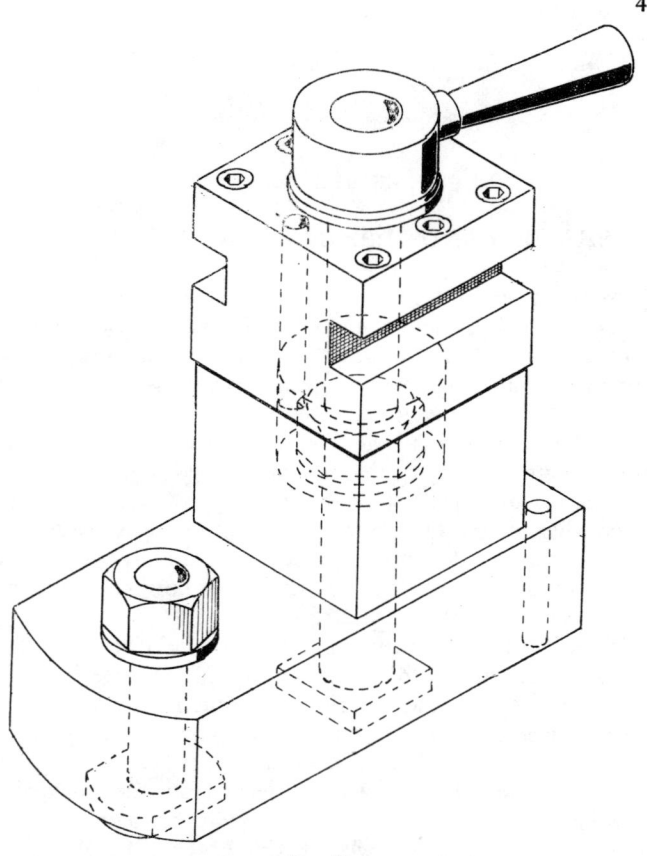

Fig. 43

at the back of the cross slide, and the two-station turret enables either tool to be brought into use at will. When mounted in position, the tool post does not interfere in any way with ordinary turning operations and it can, therefore, be kept in place ready for immediate use.

The small turret is made for holding ground, high-speed steel tools of $\frac{1}{4}$ in. square section.

CHAPTER FOUR

SHAPING AND PLANING MACHINE TOOLS

BEFORE describing in detail the tools commonly used in shaping and planing machines it will be advisable to consider briefly the conditions under which they work, bearing in mind that the two machines are essentially similar in their mode of operation.

In the shaping machine the movements of the tool are guided across the work by the ram slideways, whereas the planing machine has a stationary tool against which the work is fed when attached to a carriage moving on the bed slideways. As in either case the cutting process is in a straight line along a flat surface, it follows that the tools normally used are similar in all essential respects.

When turning cylindrical parts in the lathe, the effect of spring in the tool is to disengage the cutting edge from the work surface as shown in Fig. 44; but in the shaping or planing machine, on the other hand, any springing of the ordinary type of tool will, as illustrated in Fig. 45, tend to force the point of the tool more deeply into the work, thus possibly causing a dig-in with consequent marring of the work surface. For this reason, when taking heavy or roughing cuts it may be preferable to use a tool of the form illustrated in Fig. 46 that has its cutting edge lying on or behind the line which passes through the point of support S; in the event of the tool springing under load, it will then rise clear of the work surface along the path AB.

In tools of this pattern it is essential that the bow or cranked portion of the shank should be resistant to bending stresses, otherwise, when the cutting edge meets with an obstruction,

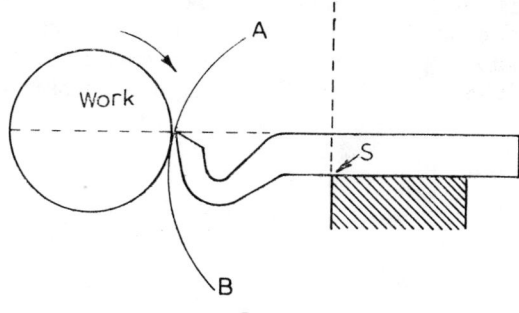

Fig. 44

the bow will open out and thus cause the point of the tool to dig farther into the work as represented in Fig. 47. To illustrate the examples cited, the dimensions of the cranked portion of the tool have been purposely exaggerated in the drawings, and in practice these tools are forged more nearly to the shape shown in Fig. 48, where it will be seen that both the shank and the head of the tool are of extremely robust construction.

Where the cranked portion is weak it is better to form the tool with the bowed portion pointing forwards as represented in Fig. 49, where a spring finishing-tool is shown ; it will be clear that here, when the cutting edge encounters any great resistance, relief will be afforded by the bow closing and lifting the point of the tool away from the surface of the work.

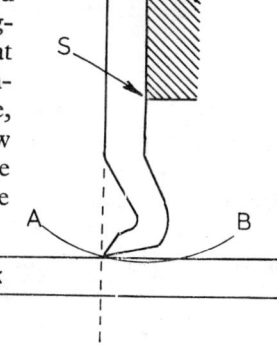

Fig. 45

Although tools of the forms described will be found to give good service in the small workshop, for commercial use the trend is to employ tools of increased cross-sectional area, mounted with a minimum of overhang in a machine of great rigidity.

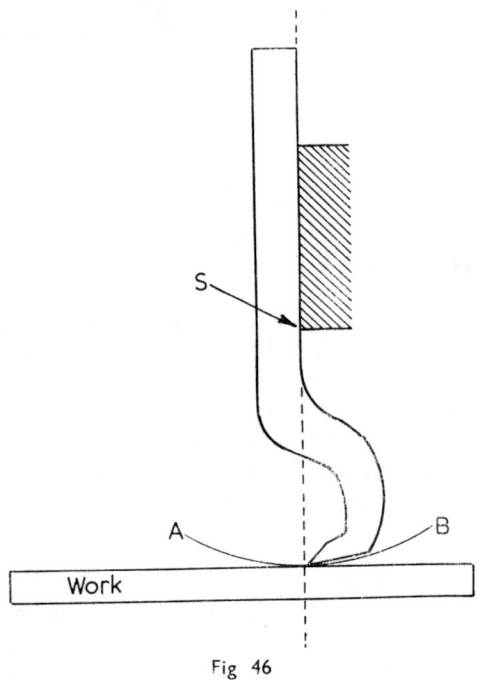

Fig 46

The tool illustrated in Fig. 50 was designed for use in a Drummond hand-operated shaping machine as a general purpose tool to save having to make a number of special tools for plane and angular shaping. As will be seen, the shank (1) carries the revolving tool head (3) well behind the line of spring, and the cutter bit is formed from a short length of ¼ in. diameter round, high-speed steel which can readily be ground to any form required.

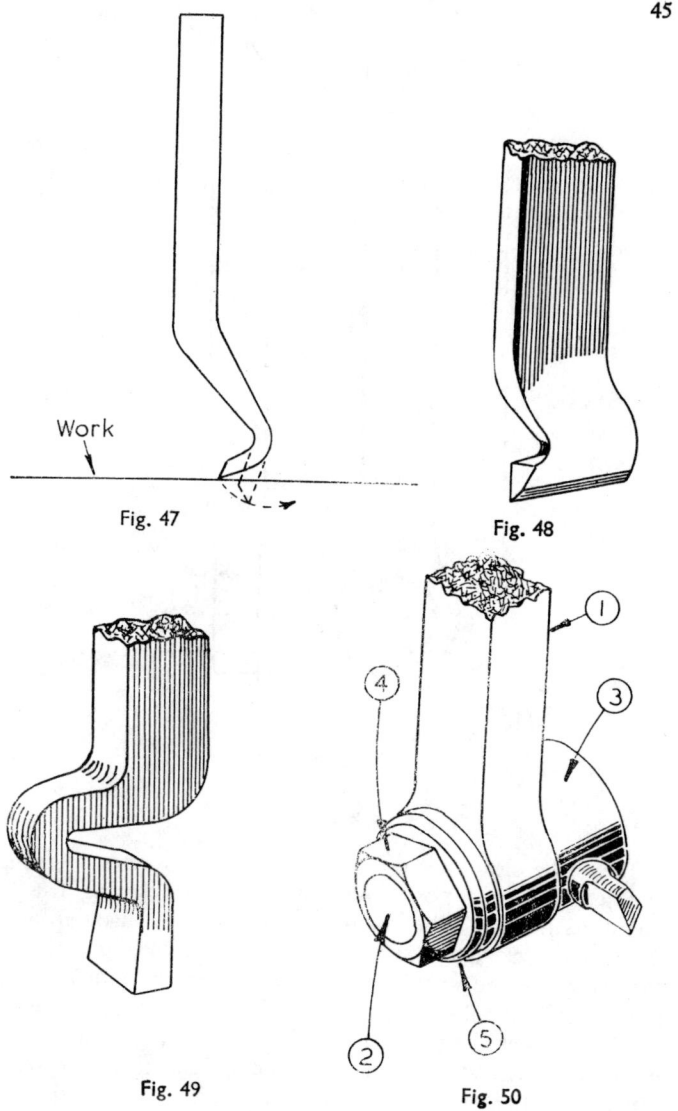

Fig. 47

Fig. 48

Fig. 49

Fig. 50

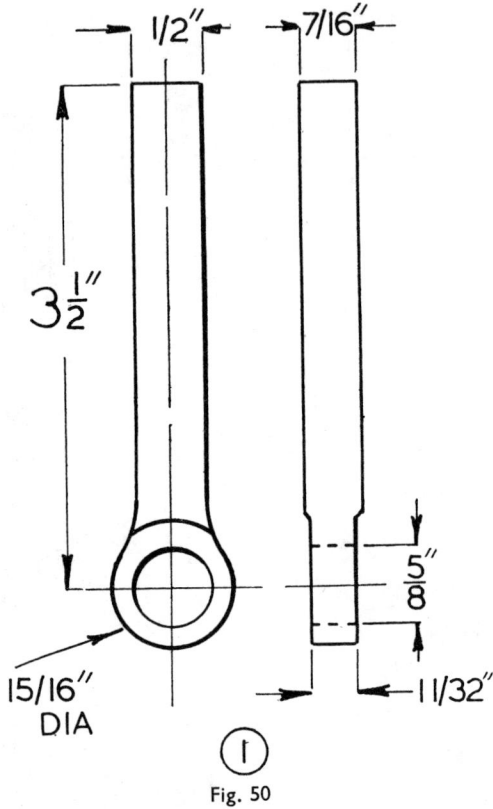

Fig. 50

It will be clear that the tool head can be rotated to bring the tool into any position necessary for machining flat or angular surfaces. The shank of the tool was made from a discarded cycle crank which, as illustrated in Fig. 51, was first shortened to enable it to be swung in the lathe, and a bolting hole was also drilled a short distance from the cut end. The crank was then secured to the faceplate, as represented in the drawing, and, at the same time, packing pieces were fitted to set the work some ¼ in. away from and parallel with the surface of the faceplate. When the pedal pin hole had been set to run truly, it was bored out to $\frac{5}{8}$ in. diameter; the front bolting face to engage the tool head was then faced, and by using a back-facing tool the inner surface was also faced true.

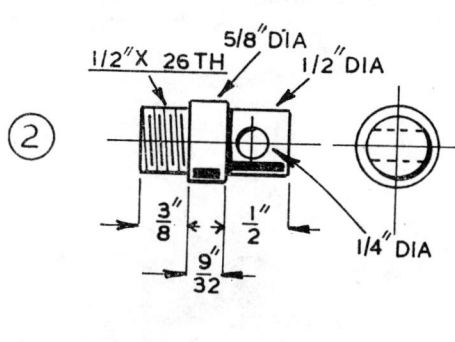

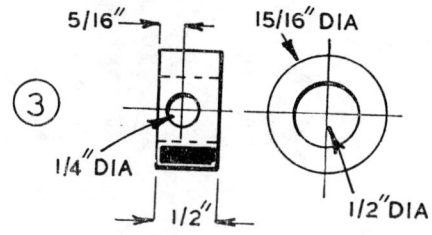

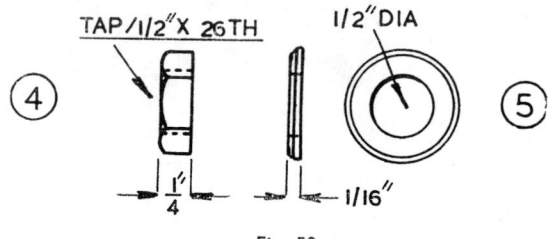

Fig. 50

To complete the shank, it was cut to length and all its four sides were filed parallel and flat.

The tool head (3) was turned from mild steel to the dimensions shown in the drawing so as to make it a good fit in the shank.

The draw-bolt (2) is threaded $\frac{1}{2}$ in. 26 t.p.i. and turned to fit the bore of the tool head ; it is advisable, in the first place,

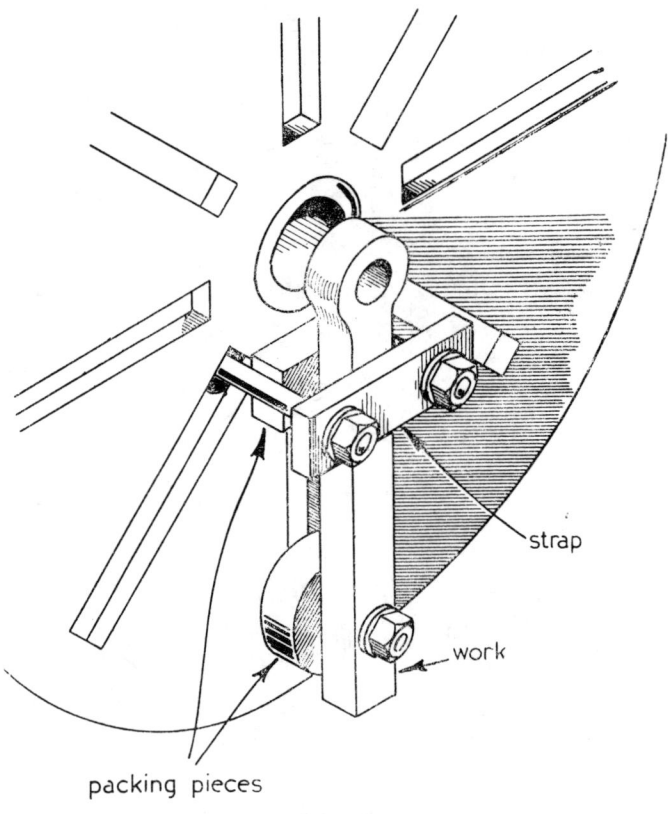

Fig. 51

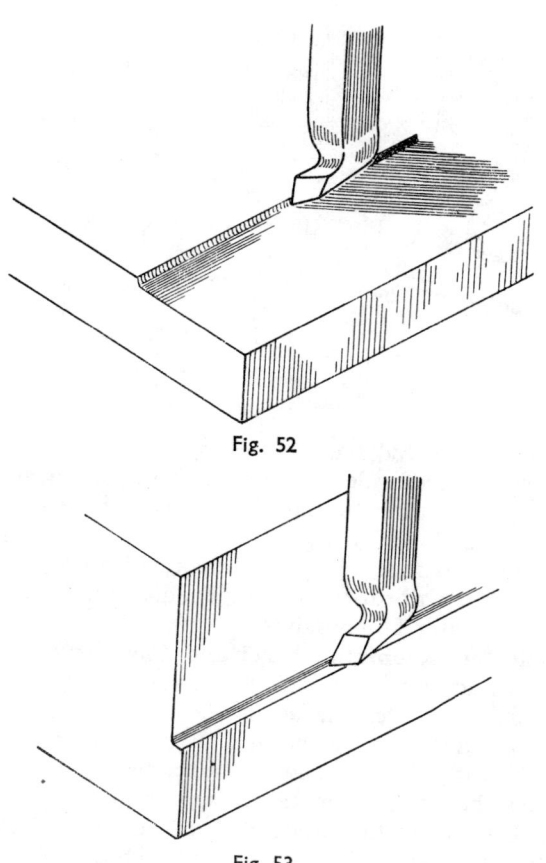

Fig. 52

Fig. 53

to make this bolt a firm press-fit so that it will remain in position during the cross-drilling operation which next follows.

This hole is reamed to size to accommodate the tool bit, and the bolt is then withdrawn and treated with a strip of emery cloth to make it a firm sliding fit in the tool head.

The nut was turned from a length of standard Whitworth

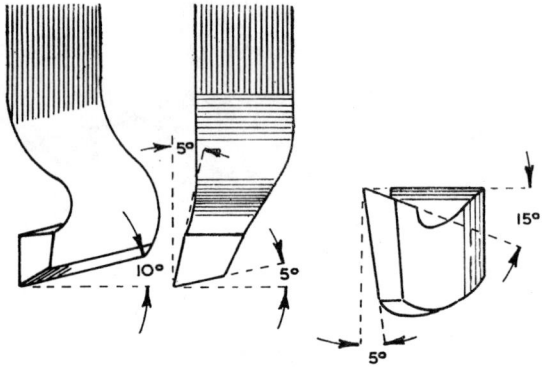

Fig. 54

nut-size hexagon rod, but a ⅜ in. hexagon nut can, if preferred, be bored and rethreaded to serve this purpose. Finally, after the nut, the draw-bolt, and the tool head have been case-hardened, the holder is assembled and is ready for use.

The tool bits should be ground in accordance with the tool forms later described, and it will be found that most of these can quite readily be reproduced.

Tools for Machining Steel and Cast Iron. The tool illustrated in Fig. 48 is suitable for machining either the horizontal or the vertical face of the work as is depicted in Figs. 52 and 53, for, as shown in Fig. 54, the tool point is provided with both top and side rake, and in addition, the clearance angles are kept small in order to maintain the strength of the cutting edges. The rake angles are essentially similar to those used for turning steel, but their actual value is, perhaps, best determined experimentally for, if they are excessive, the tool will tend to dig into the work, particularly where the machine is lacking in rigidity.

For finishing the surface of the work, either the tool point is made more rounded to afford a greater area of contact as shown

Fig. 55

in Fig. 55, or a tool with a broad, slightly curved edge as depicted in Fig. 56 can be employed for machining horizontal surfaces.

A tool rather similar in form to a lathe knife tool, as shown in Fig. 57, can be used for taking both roughing and finishing cuts, but for the latter purpose the tip of the tool should be slightly rounded where it comes into contact with the work.

Tools for Brass. As in the case of turning tools, little or no top or side rake should be given to tools used for shaping brass, but, apart from this, their form can quite well be similar to that of the tools described for machining steel.

A simple pattern of front tool suitable for shaping brass is shown in Fig. 58, but for taking finishing cuts a broader cutting edge can be employed as illustrated in Fig. 59.

Machining Aluminium. Tools intended for shaping and planing aluminium alloys have the same form as those used for machining steel, except that the rake angles can, with advantage, be increased to promote free-cutting by affording the tool a more pronounced slicing action.

From the above short account of the shaping tools in common use in the small workshop it should be clear that the tool holder with the rotating head previously described will, with

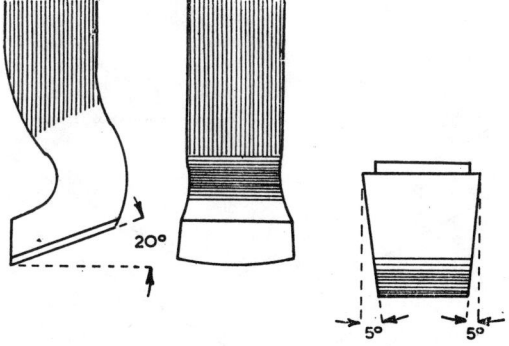

Fig. 56

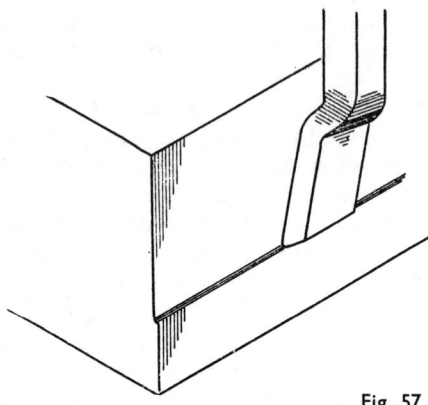

Fig. 57

the aid of suitably ground cutter bits, be able to carry out all ordinary shaping operations, for, by altering the position of the tool head, the cutter can readily be brought to bear on either a vertical or a horizontal work face.

The short lengths of ground, high-speed steel of spuare-

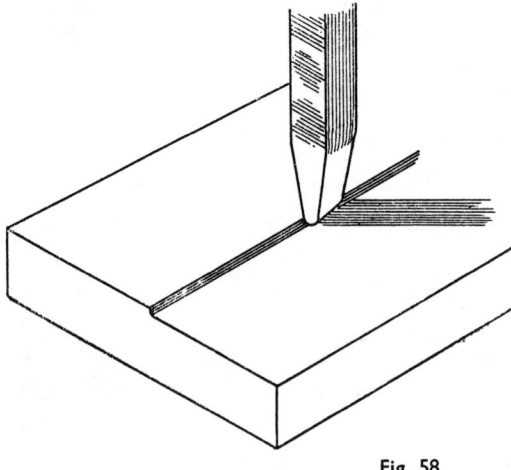

Fig. 58

section will, as in the case of lathe tools, be found most useful for making tools for use in the shaping and planing machine, and, in this connection, steel of the Eclipse brand is obtainable in sections especially adapted for this purpose.

For machining iron castings, tungsten carbide tools will cut through the hard surface scale without becoming blunted, but great care must be taken not to subject the cutting edge to

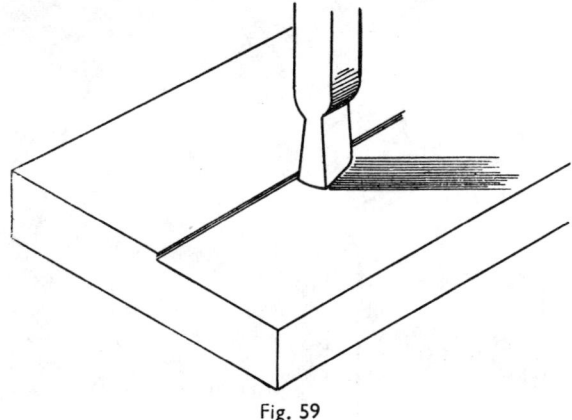

Fig. 59

shock as the brittle material forming the tip is easily damaged by sudden or intermittent stress.

Tools Used for Special Operations. A not uncommon operation undertaken in the shaping machine is the machining of V-slides, and for this purpose the tool illustrated in Fig. 48 when used in both the right- and left-handed forms will be found quite satisfactory, provided that the point is correctly shaped to clear the work faces and to work into the angle of the V.

The ram head, as illustrated in Fig. 60, is set over to provide the angular feed required, and, when an under-cutting operation such as this is undertaken, the clapper box carrying the tool must be locked to prevent the tool from rising on the return stroke and fouling the overhanging surface of the work.

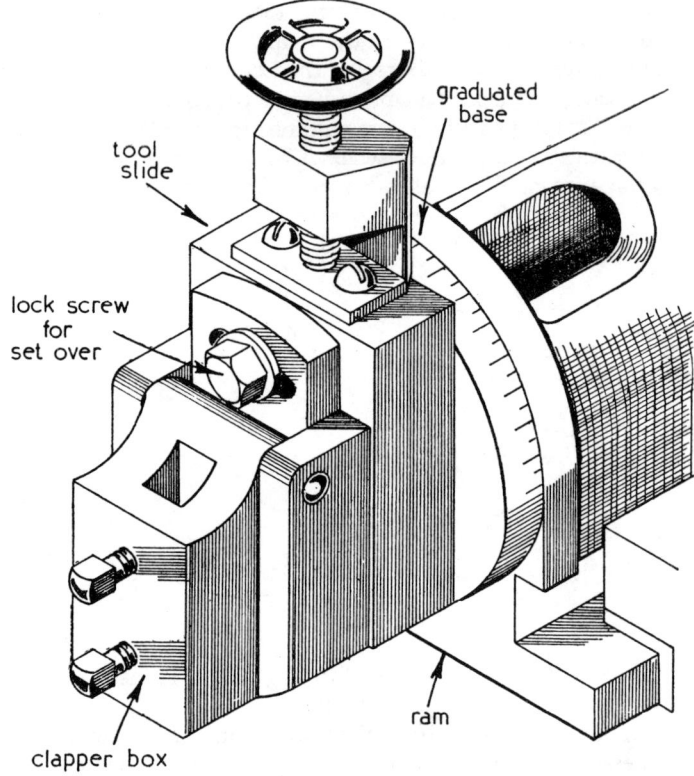

Fig. 60

A convenient and easily made form of clamp for securing the clapper box is shown in Fig. 61.

For under-cutting the sides of T-slots in a machine table, as illustrated in Fig. 62, the special form of tool depicted is required. This resembles a cranked form of parting tool, but ample clearance angles behind the cutting edge must be provided to ensure that the tool does not rub against the work.

As in the previous example, the clapper box must be

locked to prevent the tool from jamming in the work on the return stroke.

Keyways can readily be cut in shafts mounted in the shaping machine by using a tool similar in form to the lathe parting tool. When cutting an internal keyway in a sprocket or other

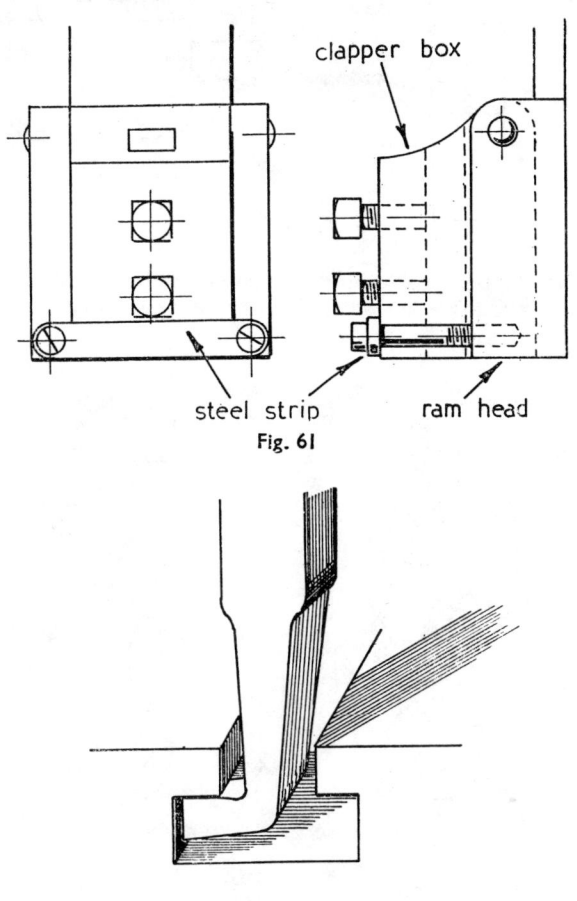

Fig. 61

Fig. 62

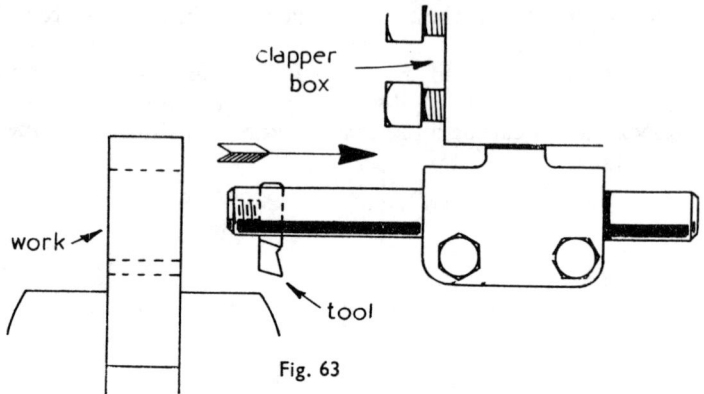

Fig. 63

The clapper box should be secured as in Fig. 61 when cutting a keyway and using reverse motion.

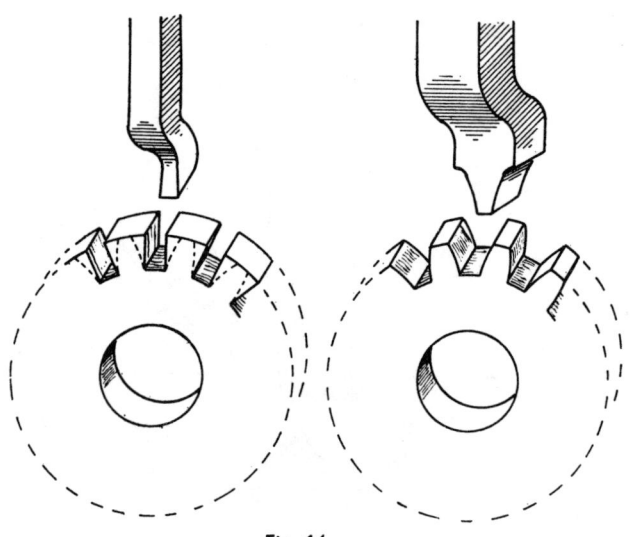

Fig. 64

part mounted on the machine table, a tool holder of the pattern illustrated in Fig. 22, Chapter Three, can be used to hold a boring bar such as the Nulok. A tool bit shaped like a small parting tool is then mounted in the bar, and, as shown in

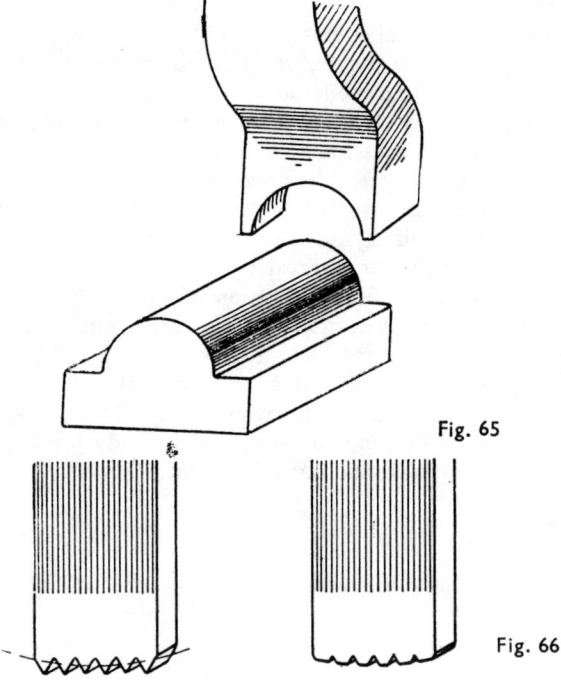

Fig. 65

Fig. 66

Fig. 63, it is employed to cut the keyway by drawing and not pushing it through the work.

Form Tools. Form-tools can be used in the shaping machine for doing a large variety of work. Pinions can be machined, with the aid of a suitable dividing gear, by first roughing out the teeth with a square-ended tool and then finishing them to the correct shape with a form-tool as represented in Fig. 64.

Fig. 65 shows a method of machining mouldings with a special form-tool, but in this case the work should first be roughed out nearly to shape, as it is possible to take only light cuts with a tool such as this, having a very broad cutting edge.

A rather unusual operation undertaken on one occasion was to machine in the shaping machine the ribbed tread-plate for the cabin floor of a model locomotive. This work was carried out with the aid of a form-tool made by grinding down the teeth of an external hand-chasing tool of 1/10 in. pitch, as illustrated in Fig. 66. The finished tool, when mounted in the shaping machine, was fed downwards until the surface of the work had acquired the correct form; the tool was then traversed the proper interval by means of the 1/10 in. pitch feed screw, and a further portion of the work was similarly machined. This operation was continued until the whole of the tread-plate was machined to a well-finished ribbed surface.

As previously mentioned, special tools and particularly form-tools, can conveniently be made from silver steel, for this material is easily filed to shape and is ready for use after it has been hardened and suitably tempered.

CHAPTER FIVE

SHARPENING LATHE AND SHAPING MACHINE TOOLS

TOOL WEAR. After a tool has been in use for some time and it is found that it will no longer produce a good finish on the work, or if abnormally high feed pressure is required to enable it to cut, then the cutting edge should be carefully examined, preferably with the aid of a lens.

Blunting of the tool, apart from actual breakage of the cutting edge, is usually the result of the clearance angles becoming reduced by wear; and if a worn tool, such as a knife tool, is critically examined it will probably be found that the point has assumed the appearance shown in an exaggerated form in Fig. 67, where it will be seen that both the side and front clearance have become obliterated for some distance below the cutting edge.

Needless to say, a tool in this advanced state of wear will merely rub against the surface of the work, and will produce a rough and ragged finish when forced to cut by the application of excessive feed pressure; further, the abnormal stresses to which both the machine and the work are subjected will result in the production of inaccurate work.

Wear of the tool will also take place at the rake surface; that is to say the area on which the chips impinge as they are separated from the work surface by the wedging action of the tool.

The latter form of wear, as illustrated in Fig. 68, appears as pitting or cratering of the tool's surface; this increases the friction required for the separation of the chips as they slide over the rake surface. This cratering effect may result in the chips adhering to the point of the tool and forming a false

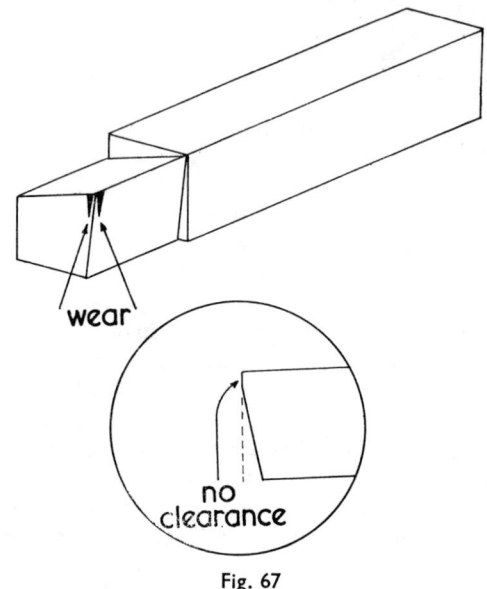

Fig. 67

cutting edge as previously described; this in itself will greatly impair the quality of the finish imparted to the work.

Cratering also undermines and weakens the cutting edge, so that in time it may lead to the actual breakage of the tool's edge at this point.

After the tool's cutting edges have been carefully examined, the next step is to decide exactly how best they can be restored to their original form by grinding. Clearly, if the upper surface of a knife tool were ground down sufficiently, the worn areas would be obliterated and sharp cutting edges again formed; however, this would not only involve the removal of much metal, but would at the same time greatly reduce the height of the tool, thus causing it to be worn out prematurely.

As a rule, the pitted or otherwise worn rake surface of the

tool requires only light grinding to form an even surface up to the cutting edge, and the bulk of the grinding, therefore, consists in reforming the clearance angles. Nevertheless, each case should be treated on its merits, so that the grinding is carried out to remove metal from situations where it does not unduly shorten the working life of the tool. The knife tool should, therefore, be ground, as far as possible, in the direction of its long axis, whereas the swan-neck pattern, used largely in the shaping machine, is ground preferably on the upper surface of the tip, in order not to reduce unduly the strength of the cutting portion.

Grinding Equipment. For grinding the tools accurately to the various forms that have been described, it is essential that the grinding machine should be furnished with a true-running wheel of the type recommended by the makers for general use in the small workshop, where the wheel will, as a rule, be used dry and not with a continuous supply of water. In addition, the wheel spindle should be accurately fitted in its bearings and should run without end-play.

At the outset, it should be emphasised, that, except in the case of the highly-skilled operator, a good result can hardly be expected if the angular surfaces forming the tool's cutting

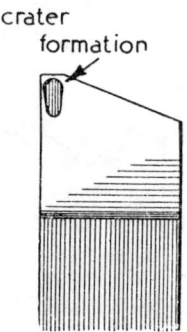

Fig. 63

edge are ground by eye ; and it is strongly recommended that any form of grinding rest employed should be so designed that the operator is enabled to grind on his tools the exact angles desired.

Although the practice of grinding the tool on the periphery

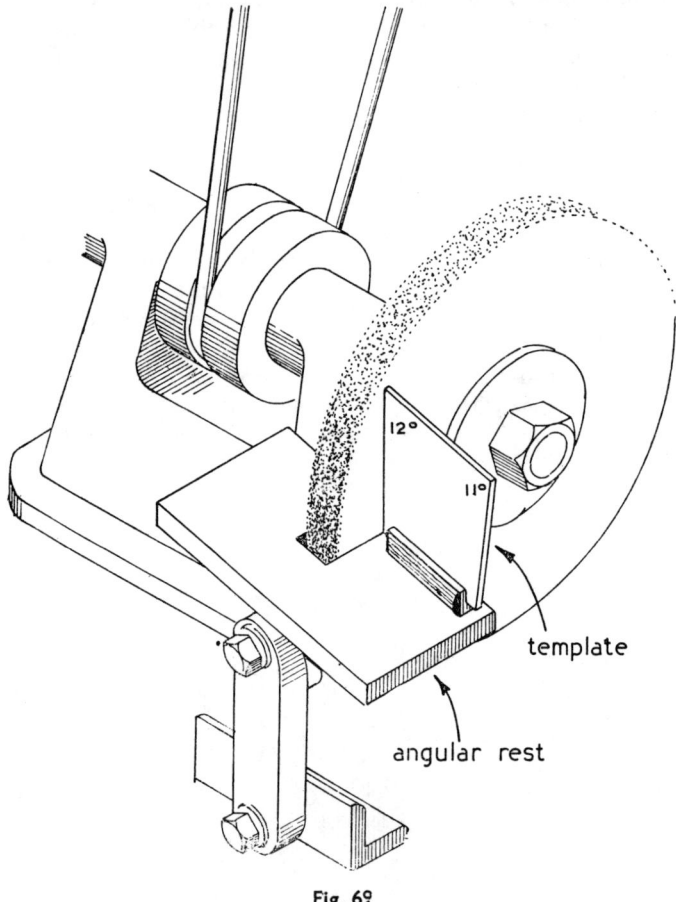

Fig. 69

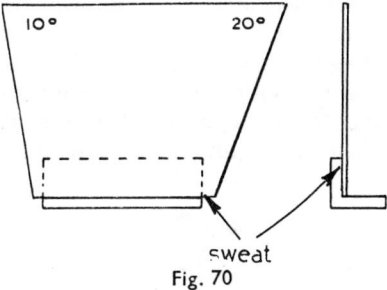

Fig. 70

of the wheel has many adherents, this procedure has the disadvantages that it is difficult to set the rest to reproduce the precise angle required on the tool ; also the tool surfaces are rendered markedly hollow-ground when a wheel of small diameter is used, thus weakening the cutting edge so formed.

If, on the other hand, the sides of the wheel are used, it is a simple matter to set a suitable form of rest to reproduce any angle required.

Angular Grinding Rest. The rest itself can either be incorporated in the grinder, if the construction of the machine allows, or the grinding rest can be secured in position to the bench top.

An angular grinding rest of simple construction and designed for attachment to the bench is illustrated in Fig. 69.

Here, it will be seen that the work table can be adjusted on its upper pivot bolt to any required angle, and, at the same time, the gap in the table is set to clear the sides of the wheel by swinging the vertical pillar on the lower pivot bolt.

Although the table can be provided with a graduated setting scale, the readings will be upset when the vertical pillar is moved sideways, and it is, on the whole, simpler and more accurate to adjust the angularity of the table in relation to the wheel by using some form of template or setting gauge. The readily constructed form of template appearing in Fig. 69 is shown in greater detail in Fig. 70, where it will be seen that a piece of sheet copper or brass is sweated to a short length of brass angle material to form a base on which the gauge can stand upright ; the template is cut and filed to the exact angle required with the aid of a protractor, and on completion the appropriate angles are stamped at the corners. Setting the rest will be facilitated if a set of templates is made to cover the range of angles normally ground on the tools.

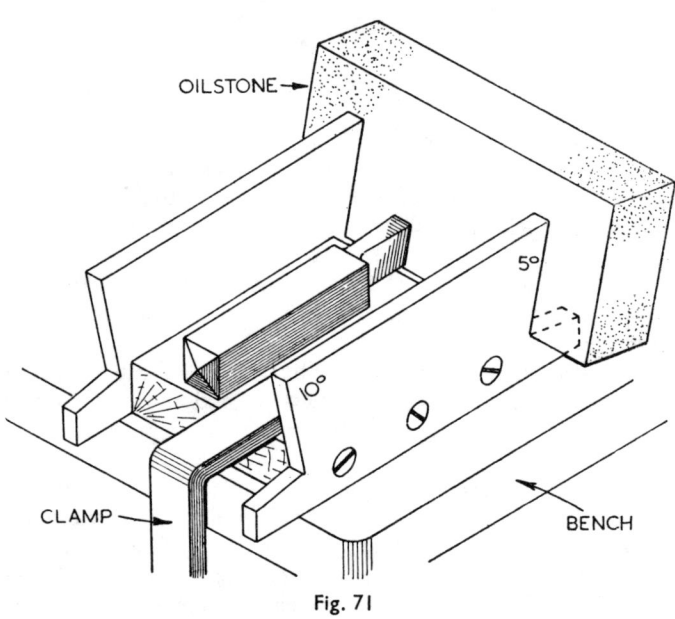

Fig. 71

In the small workshop a single wheel of medium coarseness may be found sufficient for all ordinary tool sharpening ; but, where tools have to be formed to shape and much metal has to be removed, it will be advisable to use a second wheel of coarser grade for this purpose.

Honing the Tool. The final sharpening of the end surfaces of the tool, as well as any light resharpening, is best carried out with the aid of a honing jig of the type illustrated in Fig. 71.

The steel side-plates, which are attached to a wooden base, act as guides for the Arkansas or India oilstone as it is moved to and fro in a straight path across the point of the tool, which is, meanwhile, held firmly in place with the fingers.

These side-plates may conveniently be formed to guide angles of 5 and 10 deg. at their two ends, for the smaller angle

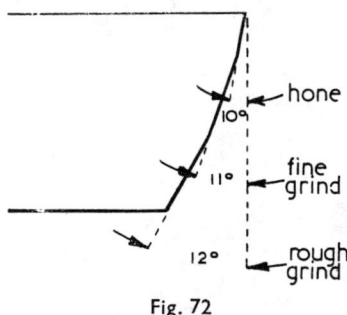

Fig. 72

will be found appropriate for sharpening parting tools. Moreover, the side-plate can be used to position the tool so that its front edge is stoned truly at right angles with the long axis of the shank.

Should the edge of the tool be sharpened with an oilstone slip held in the hand, great care must be taken to avoid rounding the cutting edge and so impairing its efficiency.

Angular Grinding. Although the coarse grinding wheel will remove metal from the point of the tool comparatively quickly, the fine wheel will cut more slowly, and the oilstone least of all; for this reason it is advisable to apportion the sharpening process in accordance with the diagrammatic drawing shown in Fig. 72.

Here, it will be seen that the rest or guide is set so that the coarse wheel grinds at an angle of 12 deg.; the fine wheel at 11 deg., and the oilstone at 10 deg. In this way, the amount of

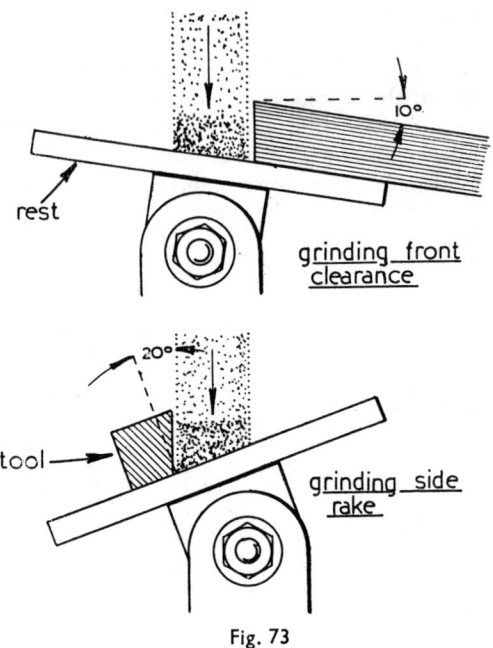

Fig. 73

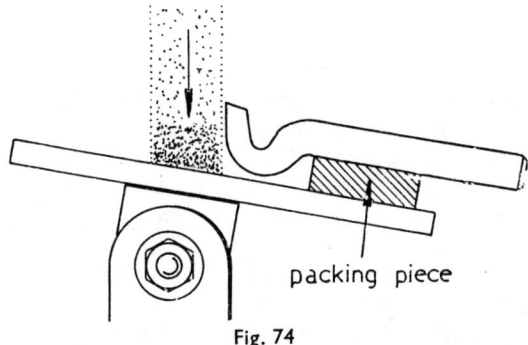

Fig. 74

metal removed subsequent to the initial rough grinding operation is small, thereby hastening the sharpening process and, at the same time, reducing the danger of overheating the tool. The values of the angles given are for the purpose of illustration of the method and should, of course, be varied to suit any particular form of tool.

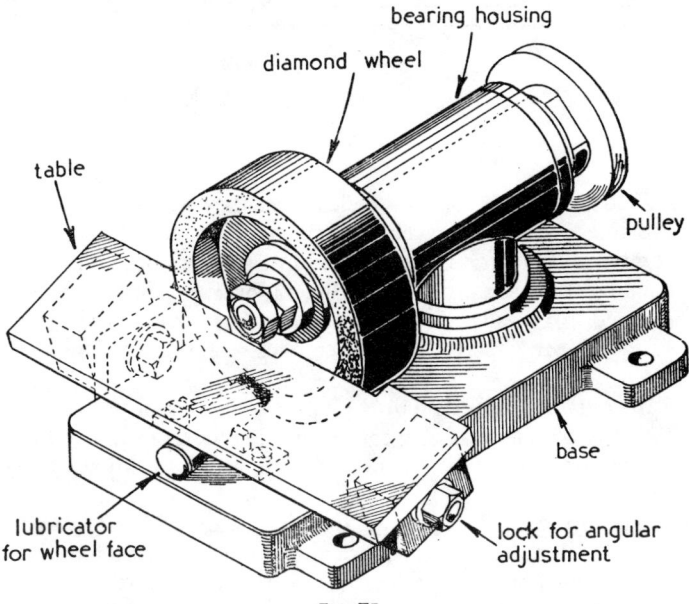

Fig. 75

The procedure outlined, and illustrated in Fig. 73, applies to the grinding of the clearance angles at the point of a right-hand knife tool formed from square-section steel; when its side rake angle is ground, the rest is set accordingly with the tool applied to the left side of the wheel as it faces the operator; this is to allow the wheel to rotate towards the cutting edge, as it always should do, and not away from it.

Where tools of irregular shape, such as a swan-neck tool,

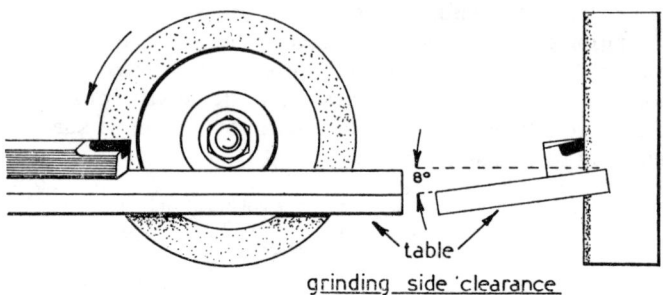

Fig. 76

have to be ground, it will be clear that they will not, as in the previous instance, lie evenly on the grinding table. To overcome this difficulty, the tool should be supported on a parallel packing strip, as illustrated in Fig. 74, or, better still, it can be gripped in a small toolmaker's clamp vice so that a flat bearing surface is provided for making contact with the work table.

Tungsten Carbide Tools. These tools cannot properly be sharpened on an ordinary carborundum wheel, and, instead, a special green-grit wheel is used for the rough grinding operation.

To fit the edge so formed for use, it can be finish-ground on a wheel impregnated with diamond dust, in order to render it sufficiently smooth and regular to enable the brittle material of the tip to withstand the stresses arising during machining operations. As these diamond wheels are necessarily expensive, it is probable that a wheel of no more than 3 in. diameter will be installed in the small machine shop; and as the manufacturers recommend that a wheel of this size should be driven at 7,000 r.p.m., a grinding head fitted with ball bearings may be preferred to a plain bearing machine. Moreover, tungsten carbide is so lacking in strength that the cutting edges of the tool should on no account be weakened by hollow-grinding the tip on the periphery of the wheel. The flat face of the wheel is, therefore, used for all grinding operations, and the most economical form of wheel for this purpose is the cup-shaped wheel carrying a band of diamond compound on its face.

In the small workshop, on the other hand, tungsten carbide tools can be finish-ground to a keen cutting edge by using a green-grit cup wheel, of 240 grain size, to follow the rough-grinding operation carried out on the 60-80 grain size, flat, green-grit wheel.

Grinding will, however, be much slower than with a diamond wheel.

If the fine-grit cup wheel is fitted to the grinding machine illustrated in Fig. 75, there should be no difficulty in giving a high finish to the tool's cutting edges and, besides preventing overheating, the supply of thin oil to the wheel will check any tendency for the grinding face to become loaded with metal particles.

Before grinding tipped tools, the mild-steel shank may need to be filed back to keep it clear of the wheel face, as this soft material would clog the wheel.

Green-grit hand slips are obtainable for honing the edges of the tool after finish-grinding. But, as these stones are of rather soft texture, care must be taken when applying them not to round the tool's cutting edges.

A Ball-bearing Grinding Machine with Cupped Wheel. This type of wheel necessitates using a hinged form of angular grinding rest, fitted to lie parallel with the surface of the wheel as illustrated in Fig. 75.

When grinding the front or side clearance of the tool, as shown in Fig. 76, the rest is tilted downwards to the appropriate angle and the direction of rotation of the wheel must be towards the edge being ground.

For grinding the side rake of a knife tool, for example, the rest is reset to the required angle and the tool is applied to the opposite side of the wheel, as illustrated in Fig. 77, in order to maintain the direction of cut towards the tool's edge.

The reversal of the direction of rotation of the wheel in this instance is effected either by means of a reversing switch fitted to the driving motor, or by the simple expedient of crossing the driving belt.

When grinding tungsten carbide tools on a wheel of this type having a resinoid base, a constant supply of thin oil is fed to the abrasive surface by means of a felt wick dipping into an oil reservoir and pressed against the wheel by a light spring.

The final step in the sharpening operation, in the case of

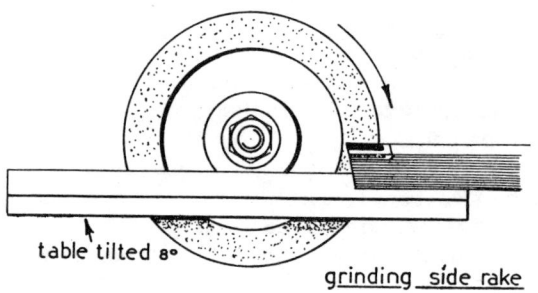

Fig. 77

tools used for machining steel, is to smooth the cutting edge with a hand lap impregnated with diamond compound.

The grinding machine illustrated in Fig. 75 has a spindle running in angular contact ball bearings, which are pre-loaded when adjusted in order to eliminate any trace of end-play.

The grinding table is hinged on two pivot bolts, and the supporting brackets are slotted to allow the table to be set close to the wheel surface when angular adjustments are made.

Both the grinding head and the table mounting are securely bolted to the heavy cast-iron base, in order to ensure rigidity and freedom from vibration when the machine is in operation.

This grinding machine can, if desired, be used for sharpening ordinary high-speed steel tools and, when used for this purpose, a less expensive carborundum wheel of either the flat or cupped form can be fitted.

Top: A carbide-tipped tool grinder, similar to that shown in Fig. 75 on page 67. Above: Lubricator for the grinding wheel as shown in Fig. 75 on page 67.

INDEX

A
Acme Threads, 33.
Aluminium Machining, 7, 14.
Angle, Clearance, 8, 20, 22, 36, 63-64.
,, Cutting, 9.
,, Front Clearance, 36.
,, Rake, 8, 11, 17, 61
,, Relief, 11.
,, Top Rake, 36.
Angular Grinding 65-68.
,, ,, Rest, 63, 69.
,, ,, ,, Template for the, 63-64.

B
Back Facing Tool, 23.
,, Tool Post, 23, 40-41.
Ball-bearing Grinding Machine, 69-70.
Boring Bar, 26.
,, ,, Bits, 24.
,, ,, Nulok, 25, 31, 57.
,, Tools, 23-26.
Brass, Machining, 13.
Bronze, ,, 13.

C
Carbon Steel, 1, 2, 6.
,, ,, Hardening, 1.
,, ,, Tempering, 1-2.
Cast Iron, Machining, 12.
Chaser, Hand, 34.
Clapper Box, 54.
,, ,, Clamp for, 54.
Clearance Angles, 8, 12, 20, 22, 36.
Copper, Machining, 15.
Cratering, 59-60.
Cross Slide, 39.
Cutting Angle, 9.
,, Speeds, 6.

D
Design of Shaping Machine Tools, 42-44.
Diamond Tools, 7, 15.
Duralumin, Machining, 14.

F
Faulty Mounting of Tools, 35.
Forged Tools, 16.
Form Tools, 2, 57-58.
Four-tool Turret, 40.
Front Clearance Angle, 9, 12, 36.
,, Rake, 10.
,, Tool, 19-20.

G
Graver, 33.
Grinding Equipment, 61-65.
,, Machine, Ball-bearing, 69-70.
,, Rest, Angular, 63-69.
,, ,, ,, Template, 63-64.

H
Hand Chaser, 34.
,, Turning Tools, 33-34.
Hardening Steel, 1.
Helix Angle, 27, 33.
High Speed Steel, 3, 6, 45.
,, ,, ,, Eclipse, 3, 16, 17.
,, ,, ,, Hardening, 3.
Holders, Tool, 17.
Honing, 64-65.
,, Jig, 65.
,, Tungsten Carbide Tools, 68-69.

K
Knife Tool, 8, 20, 59-60, 69.

L
Lathe Tool Holders, 17, 18, 24.
,, ,, ,, Types of, 36-41.
,, ,, ,, American, 36.
,, ,, ,, Boley, 38.
,, ,, ,, Drummond, 39.
,, ,, ,, English Pattern, 38.
L-Headed Tool, 21.
Lathe Tools, 16-35.

INDEX

M

Machining Aluminium, 14, 51.
,, Brass Alloys, 13, 14, 51.
,, Cast-iron, 12-13, 50-51.
,, Copper, 15.
,, Plastic Material, 16.
,, Steel, 11-12, 50-51.
,, Stainless Steel, 12.
,, Tool Steel, 12.
Mounting Tools in the Lathe, 34-41.

N

Negative Top Rake, 13.
Nulok Boring Bar, 25, 31, 57.

O

Oilstone, 21.
,, Slip, 65.
Oilstoning, 8, 14.
,, Jig, 65.

P

Packing Piece, 36.
,, Strips, 36, 38.
Parting Tool, 22.
,, ,, Burnerd Patent, 23.
Planing Machine, 42.
Plastic Material, Machining, 15.

R

Rake Angles, 8-11, 17, 61.
,, Side, 10, 29.
,, Top, 10, 12, 15, 21-22, 29.
Relief Angle, 11.

S

Screw Cutting Tools, 27-34.
,, ,, ,, Acme Thread, 33.
,, ,, ,, Jones and Shipman, 29-31.
,, ,, ,, Square Thread, 33.
Shaping Machine, 42.
,, ,, Drummond, 44.
Shaping Machine Tools, 43-53.
,, ,, Tools, for Aluminium, 51.
,, ,, Tools, for Brass, 51.
,, ,, Tools, for Cast Iron, 50.
,, ,, Tools, for Drummond Shaper, 44-50.
,, ,, Tools, for Keyways, 55-57.
,, ,, Tools, for Machining Steel, 50.
,, ,, Tools, for Mouldings, 58.
,, ,, Tools, for Special Operations, 53.
,, ,, Tools, for T-slots, 54-55.
,, ,, Tools, for V-slides, 53.
Sharpening Lathe and Shaping Machine Tools, 59-70.
Side Clearance Angle, 8.
,, Rake, 10-29.
Silver Steel, 2-3, 58.
Steel, Carbon, 1-3.
,, Hardening Carbon, 1.
,, ,, High Speed, 3.
,, High Speed, 3, 6, 45.
,, Machining, 11-12, 50-57.
,, Silver, 2-3, 58.
,, Stainless, 12.
,, Tempering Carbon, 1-2.
Stellite, 4, 6.
Square Threads, 33.
Swan-neck Tools, Grinding, 61, 67-68.

T

Tempering Steel, 1.
Tool, Back Facing, 23.
,, Boring, 23.
,, Burnerd Patent Parting, 23.
,, Form, 2, 57-58.
,, Front, 19-20.
,, Holders, 17.

Tool, Knife, 20, 59-60, 69.
,, Parting, 22.
,, Post, 39.
,, ,, Back, 23, 40-41.
,, Wear, 59-61.
Tools for Machining Aluminium, 14, 51.
,, ,, ,, Brass, 13-51.
,, ,, ,, Cast Iron, 12-13, 50-51.
,, ,, ,, Plastics, 15.
,, ,, ,, Steel, 11-12, 50-51.
,, Hand Turning, 33-34.
,, Screw Cutting, 27-34.
,, ,, ,, Jones and Shipman, 29-31.

Tools, Screw Cutting, Square Thread, 31-33.
,, Shaping Machine, 43-53.
Top Rake, 10, 12, 15, 22, 29.
,, ,, Angle, 12, 36.
Top Slide, 38-39.
Turret, Four-tool, 40.
,, Two-station, 41.
Tungsten Carbide, 4-5.
,, ,, Tools, 12, 13, 15, 34.
,, ,, Tools, Sharpening, 68-69.

V

V-tool, 21.